Andreas Sedlmayr

# Experimental Investigations of Deformation Pathways in Nanowires

**Schriftenreihe
des Instituts für Angewandte Materialien**

*Band 8*

Karlsruher Institut für Technologie (KIT)
Institut für Angewandte Materialien (IAM)

Eine Übersicht über alle bisher in dieser Schriftenreihe erschienenen Bände
finden Sie am Ende des Buches.

# Experimental Investigations of Deformation Pathways in Nanowires

by
Andreas Sedlmayr

Dissertation, Karlsruher Institut für Technologie (KIT)
Fakultät für Maschinenbau
Tag der mündlichen Prüfung: 12. Juni 2012


**Impressum**

Karlsruher Institut für Technologie (KIT)
KIT Scientific Publishing
Straße am Forum 2
D-76131 Karlsruhe
www.ksp.kit.edu

KIT – Universität des Landes Baden-Württemberg und
nationales Forschungszentrum in der Helmholtz-Gemeinschaft

KIT Scientific Publishing 2012
Print on Demand

ISSN 2192-9963
ISBN 978-3-86644-905-3

# Experimental Investigations of Deformation Pathways in Nanowires

Zur Erlangung des akademischen Grades
**Doktor der Ingenieurwissenschaften**
der Fakultät für Maschinenbau
Karlsruher Institut für Technologie (KIT)

genehmigte
**Dissertation**
von

Dipl. Ing. Andreas Sedlmayr
aus Pforzheim

Tag der mündlichen Prüfung:    12.06.2012
Hauptreferent:    Prof. Dr. rer. nat. Oliver Kraft
Korreferent:    Prof. Dr. rer. nat. Alexander Hartmaier

# Abstract

The theoretical strength of materials describes the stress necessary to shear entire layers of atoms on each other or to separate them. Defects such as dislocations and flaws reduce the strength and cause deformation and fracture at significantly lower stresses. If a material is made sufficiently small the physical boundary conditions change, resulting in an increase in strength, which is termed mechanical size effect. Experiments on specimens having physical dimensions of a few hundred nanometers show increasing strength, yet always being below the theoretically predicted values. This effect is of tremendous importance for the reliability of small scale building blocks, e.g. in nanotechnology. Beyond that the deformation at small scale is of fundamental scientific interest, since a widely accepted picture of the governing deformation mechanisms does not exist. Besides the physical boundary conditions, the methods of preparation and statistical effects have to be taken into account, in order to achieve a deep understanding of the peculiar behavior.

In this work nanowires were subjected to monotonic tensile stresses to investigate the strength and the accompanied deformation behavior in the size regime of 50 to 500 nm. Therefore, nanowires with different atomic bonding and high crystalline purity were deformed in the scanning electron microscope. The resulting mechanical response and the corresponding deformation were monitored and evaluated. All investigated materials show high strengths, in some cases on the lower end of predictions for the theoretical strength of the respective material. Furthermore, the tested nanowires show stochastic behavior with respect to the yield strength and ductility revealing different amounts of variation. In the virtually defect free materials this can be attributed to the fatal influence of single preexisting atomic defects or their nucleation, most likely at randomly distributed imperfections at the surface of the nanowires. Moreover at these small length scales, the borders of strict mechanical behavior of materials are fading. Changes and acceleration of deformation mechanisms are observed, e.g. the occurrence of deformation twinning in fcc Au at room temperature. The measured high strength of the different materials and the resulting mechanical response in combination with the possibility to control materials at an atomic level are disclosing new fields for applications in nanotechnology.

# Kurzfassung

Die theoretische Festigkeit von Materialien beschreibt die Spannung, die nötig ist um ganze Lagen von Atomen auf einmal gegeneinander abgleiten zu lassen oder diese voneinander zu trennen. Defekte, wie Versetzungen und Risse, mindern die Festigkeit und führen zu Versagen bei weitaus geringeren Spannungen. Wird eine Materialprobe nun hinreichend klein, so ändern sich die physikalischen Randbedingungen. Daraus resultiert ein Anstieg der Festigkeit, der als mechanischer Größeneffekt bekannt ist. Versuche an Proben mit Abmessungen von wenigen hundert Nanometern zeigen ansteigende Festigkeiten, jedoch immer noch unterhalb der theoretisch bestimmten Werte. Dieser Effekt ist von großer Bedeutung für die Zuverlässigkeit von kleinskaligen Bauteilen, z.B. in der Nanotechnologie. Darüber hinaus ist die Verformung in kleinen Skalen von grundlegendem wissenschaftlichem Interesse, da ein breites Verständnis der zugrunde liegenden Verformungsmechanismen nicht existiert. Weiterhin müssen neben den unterschiedlichen physikalischen Randbedingungen auch Präparationsmethoden und statistische Effekte in Betracht gezogen werden, um ein tiefes Verständnis für dieses veränderte Verhalten zu erreichen

In der vorliegenden Arbeit wurden Nanodrähte unter Zug belastet, um die Festigkeit und das Verformungsverhalten im Bereich von 50 bis 500 nm zu untersuchen. Dafür wurden Nanodrähte mit unterschiedlichen atomaren Bindungsarten und hoher kristalliner Reinheit im Rasterelektronenmikroskop verformt und die resultierende mechanische Antwort als auch die optischen Beobachtungen aufgezeichnet und ausgewertet. Alle untersuchten Materialien zeigen hohe Festigkeiten, in einigen Fällen an der unteren Grenze der vorhergesagten theoretischen Festigkeit. Darüber hinaus zeigen die getesteten Materialien stochastisches Verhalten bei dem die Festigkeit und die Bruchdehnungen unterschiedlich starker Variation unterliegen. In diesen nahezu defektfreien Materialien ist dies weitgehend dem fatalen Einfluss von schon existierenden einzelnen atomaren Defekten deren Nukleation, sehr wahrscheinlich an stochastisch verteilten Imperfektionen an der Oberfläche der Nanodrähte, zuzuschreiben. In diesen Längenskalen verwischen die Grenzen von striktem mechanischem Verhalten einzelner Materialien. Es werden Wechsel und Beschleunigung von Verformungsmechanismen beobachtet, wie z.B. das Auftreten von Verformungszwillingen in kfz Au bei Raumtemperatur. Die gemessene Festigkeit der einzelnen Materialien und die Möglichkeit das resultierende mechanische Verhalten durch die Manipulation einzelner Atome zu kontrollieren, öffnen neue Felder für Anwendungen in der Nanotechnologie.

# Acknowledgements

Die vorliegende Arbeit entstand während meiner Zeit als Doktorand am Karlsruher Institut für Technologie (KIT) am Institut für angewandte Materialien (IAM). Im Folgenden möchte ich mich bei den Personen bedanken, die zum Gelingen dieser Arbeit beigetragen haben.

Herrn Prof. Dr. Oliver Kraft danke ich an erster Stelle für die Annahme als Doktorand an seinem Institut und sein stetiges Interesse an meiner Arbeit. Bedanken möchte ich mich auch für die fortwährende Bereitschaft zu Diskussionen und seinen wissenschaftlichen Input.

Herrn Prof. Dr. Hartmaier danke ich für die Übernahme des Koreferats.

Meinem Betreuer, Herrn Dr. Reiner Mönig, danke ich für die sehr gute und freundschaftliche Zusammenarbeit und seinen fachlichen Einfluss.

I would like to thank my second advisor, Prof. Daniel Gianola, for his enduring support during this work. He helped me to get started at KIT and, besides his professional input, taught me much about enthusiasm for scientific work.

I would like to thank my external collaborators, Prof. Erik Bitzek, Prof. George Pharr, Dr. Gunther Richter, Prof. Marc Legros and Kurt Johanns, for the close work and the scientific advice and discussions.

Dr. Patric Gruber, Dr. Steven Boles, Dr. Zung-Sun Choi, Dr. Martin Härtelt, Jochen Lohmiller, Simone Schendel, Di Chen und Goran Kilibarda danke ich für die gute Zusammenarbeit und hilfreiche Diskussionen auch fernab vom Thema meiner Arbeit.

Ich danke allen Kollegen am IAM-WBM für die freundliche Zusammenarbeit und das gute Arbeitsklima.

---

Meinen Eltern und meinen Brüdern danke ich für die Unterstützung auf meinem Lebensweg und dafür, dass sie mir diese Arbeit ermöglicht haben.

Meiner zukünftigen Frau Nina danke ich für ihre Liebe und ihre Unterstützung in allem was ich tue. Dir widme ich diese Arbeit.

# Contents

# 1 Introduction and Background

## 1.1 Motivation and Technological Driving Force

The term "Nanotechnology" was first introduced by Norio Taniguchi in 1974 and is nowadays defined as an area of research and application of structures with at least one dimension being in the range of one single atom to 100 nm. Indeed noble prize winner Richard Feynman was the first to envision the possibility and the manifold opportunities that would accompany a further reduction in size in his famous lecture "There is plenty of room at the bottom" in 1959 [1]. However, like in every other technology, the emergence of the field of nanotechnology strongly hinged on the advancements of new technologies and their capabilities of imaging, handling and creating these small structures. A huge step forward was achieved with the invention of the scanning tunneling microscope (STM) by Gerd Binning and Heinrich Rohrer in 1981, which allows to image surfaces at an atomic level.

The potential of nanotechnology has been realized early by the German government and resulted in funding through various projects since the early 1990's. In 1998 the Bundesministerium für Bildung und Forschung (BMBF) started a separate funding only dedicated to this field. The subgroup of nanomaterials receives its own funding since 2001. The dynamic progress in nanotechnology manifests itself in strong increases in funding, patents, publications and products on the world market. Figure 1.1A shows the funding of nanotechnology heavily increasing, which also shows the importance of this field. As a result of the invested money, many research facilities have started to work in this field, which in turn resulted in a rise of publications on nanotechnology (Figure 1.1B).

Today, nanostructures have found their way into many applications, e.g. chip technology, where the building blocks have dimensions in the two digit nanometer regime. This can on one hand be attributed to the successive miniaturization in many fields. On the other hand, if a material's dimension is brought to the nanoscale, the physical and chemical properties, such as thermodynamics or surface to volume ratio, can change dramatically. For example, the mechanical response of the same material has been found to significantly differ between nanostructures and their bulk counterparts [2].

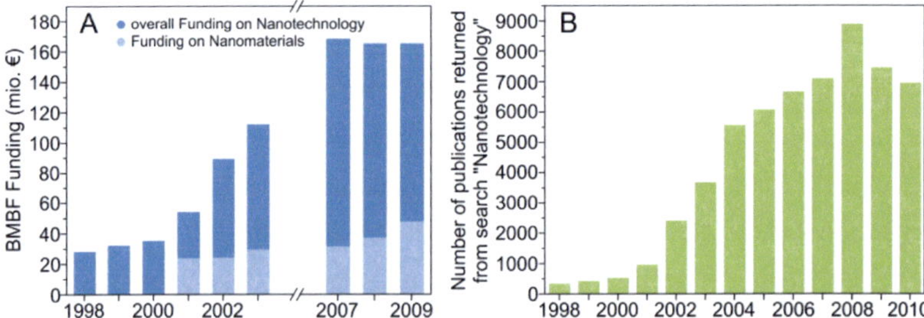


Figure 1.1: BMBF funding in the field of nanotechnology (A) since the beginning in 1998 and the number of publications returned if Scopus is searched for the term "Nanotechnology" (B), substantiating the strong increase coupled to the invested money.

If subjected to mechanical load, the former can withstand eminent larger stresses and strains. In material composites like microchips during fabrication the difference in thermal expansion coefficients between metal and semiconductor structures cause stresses, which exceed the limit of what the corresponding bulk materials do withstand. Nanostructures are suitable for sustaining the structural integrity since they can accommodate these deformations. Although the mechanisms for this behavior have not been clearly identified, nanostructured materials are regarded as promising building blocks for future devices in many applications, such as advanced

Lithium ion batteries (LIB) [3], disclosing new opportunities in the design of next generation devices. In order to exploit this "mechanical size effect", it is essential to fundamentally describe the mechanical behavior, borne from the importance of the reliability of single structures, which is key to the functionality of the overall device. Fundamental understanding of the behavior of the smallest parts is further expected to impact the theory of solids and facilitates proper tailoring and tuning of materials and components at this length scale, fortifying the need for accurate scale-specific testing methods.

Nanowires represent a special class of quasi one-dimensional (1D) nanostructures, exhibiting aspect ratios of 600 and more. They can be single- or polycrystalline. Synthesis is rational and predictable in a way that key parameters like chemical composition, diameter, length and doping can be controlled during growth [4]. Combined with superior mechanical properties, these facts have created various strategies for the integration of nanowires in future devices. For example, nanowires have been successfully used in nanoscale field-effect transistors (FET) [5], light-emitting diodes (LED) [6] or gas sensors [7]. Over the last decade, the interest in these structures has substantially grown (Figure 1.2A). Just recently, nanowires have gained access into the field of energy harvesting and storage. They are considered as cheap and efficient solar cells [8]. Furthermore, nanowires have proven to be potential electrode materials in advanced LIBs [9], where they have shown to increase the cyclic stability and therefore to affront degradation, which is one of the utmost concerns in LIBs for electromobility [10].

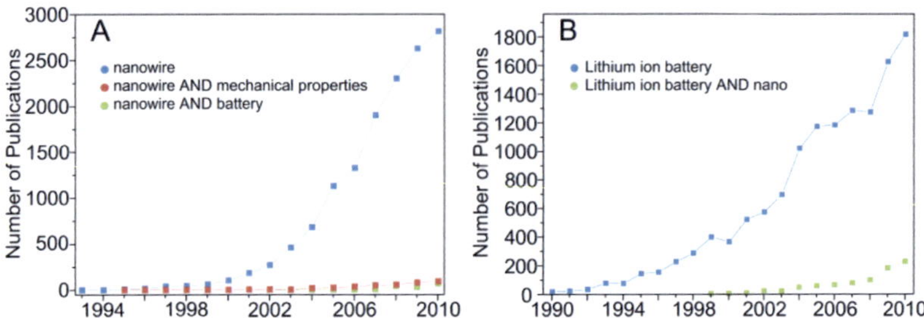

Figure 1.2: Number of publications returned from Scopus searches as written in the legends. (A) shows a strong increase in publications on nanowires, however a much lower slope is obtained on their mechanical properties and their utilization in Lithium ion batteries. The number of publications on LIBs has also strongly increased, whereas the work on nanomaterials has just started (B).

The work being presented in this thesis is motivated by various challenges. Many tests on several metallic material systems have shown an increase in strength for small materials (down to diameters of 200 nm) if subjected to compressive stresses [11], from here on referred to as "size effect". However the reason for this dramatic change in behavior, as well as the mechanical behavior below this regime still remain unclear. Furthermore, simulations predict transitions in deformation mode for various materials, if brought to nanometer dimensions, e.g. transition from ductile-to-brittle behavior or vice versa [12]. The mechanical response of nanowires and other structures below 100 nm is not easily accessible by conventional experiments. In terms of nanowires, only few experimental data on the mechanical response is available, however, data in this regime is needed to further investigate the size effect and to verify or disprove simulations. Nanowires, which are grown with high crystalline perfection in a nearly defect-free state, are perfectly suitable as model systems for materials research. By elucidating their deformation behavior it is expected to uncover the underlying deformation mechanisms in single crystals at the nanoscale.

One of the large efforts today is the progress in energy storage and electromobility. Because of their high energy density, LIBs exhibit great potential for future applications. In 2007, the BMBF has launched an innovation alliance called "Lithium Ionen Batterie LIB 2015". Within this framework, several key players among this field have committed to invest 360 millions of euro into research and development of LIBs. For the development of new, efficient systems, electrodes with maximum Li storage capacity need to be created. During cycling, the electrode material is subjected to mechanical deformation due to insertion and extraction of Li into and out of the electrode crystal lattice, leading to degradation and failure. This fact even excludes some promising materials from application. Nanoscaled particles are used in LIBs, not only because the high surface-to volume ratio is beneficial to Li diffusion and electrical conductivity, but also because the nanoscale particles can better accommodate the deformation. The knowledge about these nanomaterials is still nascent and the mechanical properties are barely investigated. Electrochemical and mechanical characterization of nanowires of LIB anode materials is considered to provide information, not only on the performance of nanowire anodes in LIBs, but also to characterize new materials for their suitability as electrode materials and primarily to understand how the mechanical properties influence the performance of a battery.

## 1.2 Thesis overview

In this thesis, the strength and the corresponding deformation behavior of quasi 1-dimensional nanomaterials, which are expected to fundamentally differ from the mechanical behavior of their bulk counterparts, are elucidated. Since no commercial experimental setup was available, two new setups for tensile testing of nanostructures were developed for these studies. The deformation behavior was then investigated in a variety of material systems having different crystal structures and distinct types of atomic bonding ranging from metallic over covalent to ionic. All results were carefully examined and discussed in the context of mechanical size effects.

The thesis is structured as follows. Chapter 1 introduces the reader into the field of nanotechnology and highlighted some outstanding work in this area. In the following sub-chapters the fundamentals, which are key to understanding of this work, are presented. Starting point are the fundamentals of deformation and fracture behavior of the materials used in this work, including Au, Si, NaCl and $VO_2$. Then an overview of established small scale mechanical experiments that have been performed in the past decades is given. Since manipulation and testing at the nanoscale is fundamentally different to the macro world, new approaches have to be taken and also standardization of testing methods is widely missing. First, requirements for proper performance of nanoscale experiments, as well as advantages and disadvantages of the presented systems and testing routes are described. Subsequently, selected results are presented addressing the "mechanical size effect" and furthermore, the propositions for the departure from conventional behavior are highlighted as they have been discussed among the community. Finally, distinct synthesis routes of Si [13], Au and NaCl [14] nanowires are illustrated.

The core work of this thesis starts out with a chapter on the experimental effort that has to be accomplished to obtain a clean mechanical test, which is readily interpretable. The construction of a mechanical setup is described. Since the specimen dimensions require image acquisition with the electron beam in high-vacuum (HV) environments and the application of forces and displacements in exceptionally small quantities, non-conventional approaches have to be used. The devices for harvesting, transfer, positioning of the nanowires as well as actuation and measurement principles are explained. With this knowledge, the two novel setups are presented and evaluated according to their performance and noisefloor characteristics. The first setup is a design iteration of a setup built by Daniel Gianola at KIT, the second setup is built from scratch. In a further section the route for proper tensile testing of a quasi-1D nanostructure is depicted and the key points for a successful conduction are explained. Afterwards the basics and application-specific conditions for digital image correlation based strain measurement are presented and discussed.

The results for tensile testing are commenced by a chapter on tensile testing of Molybdenum fibers. Specimens with different amounts of pre-strain are tested to investigate the influence of initial defect density on the mechanical response. The observed behaviour is largely scattered, due to the inhomogeneous distribution of defects, ranging from linear-elastic failure at high stresses to deterministic bulk-like behavior. In fibers having inter-mediate dislocation densities intermittent plasticity is observed. The results are compared to micropillar compression testing performed on the same material. Coming from these sub-μm sized fibers the size of the tested specimens is now decreased. In the following chapter vanadium-dioxide nanowires were tested. These nanowires show a phase change induced by mechanical straining. The corresponding phase change stress is a suitable measure to demonstrate the accuracy of the tensile tests. In the following chapters the results for tensile testing of Si <111> and <100> nanowires, NaCl nanowires and Au nanowires and nanoribbons are presented. Additional work performed to elucidate the microstructure and the coupled deformation behavior, comprising transmission electron microscopy, electron backscatter diffraction and energy dispersive X-ray spectroscopy, are shown. Si <111> nanowires show consistent behavior but changed variability for two different methods of growth. Weakest-link theory is employed to discuss the origin of the observed differences. Short Si <100> nanowires show a departure from conventional behavior of covalent bonded materials, which has only been observed in simulations, whereas long specimens show the expected brittle mechanical response. NaCl nanowires exhibit mechanical behavior, which is known but not necessarily expected in this material at room temperature. Reasons for this transition are discussed on the basis of crystal purity in conjunction with the competition between shear and cleavage due to favored cleavage and glide planes in simple ionic crystals. Despite yielding at high stresses, two classes of mechanical response and a transition in deformation behavior are found for Au nanowires. The results are discussed in the context of partial dislocation plasticity and mechanical size effects. Weibull theory is applied to account for the stochastic behavior of this material at the nanoscale. In the final chapter the key findings are revised and put into context.

## 1.3  Deformation and Fracture Mechanisms

In this chapter the conventional deformation and fracture mechanisms of the materials used in this thesis are highlighted. First, the ductile deformation of metals is discussed followed by that of brittle materials as semiconductors. Finally, the deformation of simple ionic crystals is shown. Since the work is performed on single crystals the text mostly addresses this aspect.

### *Metals*

The theoretical strength (also termed ideal strength) of a ductile material is defined as the shear stress required to slide two crystallographic adjacent planes against each other [15-17]. Figure 1.3 shows a schematic of two planes of atoms. It can be seen that the applied shear stress causes a displacement $x$ in the shear direction. The shear stress required to hold the layer in the initial position is zero, as well as the shear stress if the movement of one atom spacing $b$ is performed. Also, if the atoms of the upper layer (Figure 1.3) are halfway slid over the bottom layer the shear stress is zero. In between these points the required shear stress is a periodic function.

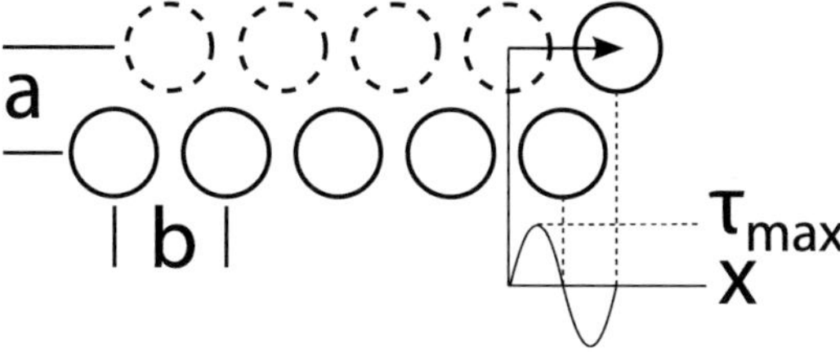


Figure 1.3: Schematic of theoretical strength of a crystalline material. In its simplest way the theoretical strength is defined as the shear stress required to move two layers of atoms relative to each other. This can be expressed by a sinusoidal relationship.

Following Dieter [15] the shear stress required to move the upper layer can be approximated by a sine wave

$$\tau = \tau_{max} \sin\frac{2\pi x}{b} \approx \tau_{max}\frac{2\pi x}{b} \tag{1}$$

where $\tau_{max}$ is the amplitude and $b$ is the period of the sine wave. For small displacements Hooke's law can be applied

$$\tau = \mu\gamma = \mu\frac{x}{a} \tag{2}$$

where $\mu$ is the shear modulus and $\gamma$ the shear strain. If the two equations are merged, the following equation provides the relationship for the maximum shear stress at which slip should take place

$$\tau = \frac{\mu}{2\pi}\frac{b}{a} \tag{3}$$

If the distances $a$ and $b$ are taken to be equal, a simple estimate of the theoretical shear strength of a perfect crystal lattice is given as

$$\tau = \frac{\mu}{2\pi} \tag{4}$$

If Gold, which has a shear modulus of 27 GPa, is taken as an example the theoretical shear strength is calculated to be roughly 4 GPa. This model only deals with the behavior inside the crystal. Effects of surfaces are not considered and also no information can be obtained whether the surface areas are intrinsically different to the inner part of the crystal.

The above derived formulas only being rough calculations, more realistic values can be obtained by computational means. Nowadays, the increased computational power facilitates atomistic simulations of the deformation of perfect crystal lattices [18]. Using ab initio density functional theory (DFT) calculations, solids can be investigated at the limit of their structural stability, to reveal the upper and lower bounds of shear at which a perfect crystal becomes unstable.

However, this high strength calculated theoretically is never observed in real materials, which is due to defects that do exist in every material causing the resulting strength to decrease to much lower values than theoretically predicted. The conventional deformation of metals is governed by dislocations, which are linear defects in the crystal lattice, mediating the plastic deformation of the corresponding material. Typical metals contain relatively large densities of dislocations (on the order of $10^{12}$-$10^{16}$ m$^{-2}$). The motion of dislocations requires much lower stresses since the lattice in distance to the defect is still close to perfect and only smaller changes near the dislocation line are necessary to initiate glide. Dislocations have extensively been discussed in the literature. For further information, exceeding the scope of this brief introduction, the reader is referred to textbooks [15-17]. Dislocations exist in almost every crystalline material. The nature of bonding between atoms in metals allows them to be more or less mobile in the atomic lattice, depending on the atomic bonds and temperature. Motion can be accomplished through glide along a crystallographic plane or through climb to a different plane, which requires diffusion and is not considered here. The motion of many dislocations results in slip, which is mainly responsible for plasticity as the governing mechanism of deformation in crystalline metals. In face-centered cubic (fcc) materials, slip is favored on specific crystal planes (slip planes) and along certain crystallographic directions, namely the planes showing highest atomic density and the directions that are closest packed. This is easy to understand if one thinks of frictional resistance. The most favorable conditions for glide would be the ones with the smoothest planes and the shortest steps.

In the fcc lattice these are {111} planes and <110> directions, together forming close packed slip systems. If redundancy of planes and directions being equal or opposite to each other is accounted for, there are 4 possible slip planes with each of them having 3 possible slip directions, leading to 12 slip systems. One glide step of a dislocation is expressed by the Burgers vector, which is a/2<110> in magnitude for fcc crystals with a being the interatomic distance. The activation of a specific slip system strongly depends on the critical resolved shear stress (CRSS), which is the shear stress acting on the slip plane in slip direction. This is rationalized through the geometry of the crystal and the orientation of the slip planes with regard to the applied stress and is commonly described by the Schmid factor $m$, geometrically relating the resolved shear stress to the applied normal stress. Hence the primary slip system is the one with the highest Schmid factor.

Since energetically more favorable, the full dislocation with magnitude a/2<110> is split into two partial dislocations in <112> directions, each of them producing slip of a/6<112>, often referred to as Shockley partial dislocations. If the atomic planes are considered as carpets of hard spheres with stacking sequence ABCABC… a full dislocation in plane B would comprise 2 sequences, first a leading partial dislocation moving the lattice plane to site C, producing a stacking fault. Second, a trailing partial dislocation completes the full dislocation motion to B. By doing this it cleans off the created stacking fault and yields a displacement of one Burgers vector.

An alternative path to such ordinary dislocation plasticity (ODP) is deformation twinning (DT). In this mechanism, competing with ODP as the carrier of plasticity, a part of the crystal is sheared by a fixed magnitude leaving two contiguous parts, the twin and the matrix (Figure 1.4). For fcc metals the crystal is transformed into a twin by successively sliding {111} planes by the amount of a/6<112> of one twinning partial dislocation relative to the adjacent plane beneath. This highly cooperative motion of twinning partial dislocations with the same Burgers vector will further extend the twin. With twinning, a portion of the crystal is transformed into a mirror symmetric position to the parent crystal about the twinning plane, which both parts of the crystal share. Figure 1.4 shows a schematic of twinning. The twinning plane is perpendicular to the paper plane. A shear stress will cause a part of the crystal to twin about this twinning plane and leave the matrix at the bottom unchanged. Although the crystal structure is maintained it is obvious that the process involves a reorientation of the twinned portion.

Deformation twinning is not expected for fcc crystals under conventional conditions, since there are many possible slip systems that would preferably accommodate deformation. DT is only the predominant mode of plastic deformation at the extreme conditions of shock loading and/or low temperatures [19]. However, plastic deformation in nanoscale volumes is markedly different to its macroscopic counterparts and twin deformation has been experimentally observed in nanocrystalline (nc) as well as single crystalline nanoscale materials [20-24] at room temperature (RT) and experimental strain-rates. The crossover in deformation mechanism has been attributed to the competition between the high stresses required to bow a full dislocation through a nanoscale grain and the energy penalty of forming stacking faults to commence twinning [20]. To propagate the twin-boundary further twinning dislocations must be stimulated on adjacent slip planes. The chance of stimulated twin propagation is strongly depending on the mechanism of twin propagation, e.g. the pole mechanism [24] or grain boundary processes [25, 26]. In contrast to nc polycrystalline materials, the observation of deformation twinning in nanoscale fcc single crystals has been mostly limited to atomistic simulations [27-32], which employ testing conditions far from experimental reality. Only recently, twinning in single crystalline gold nanowires was observed in experiments [33, 34].

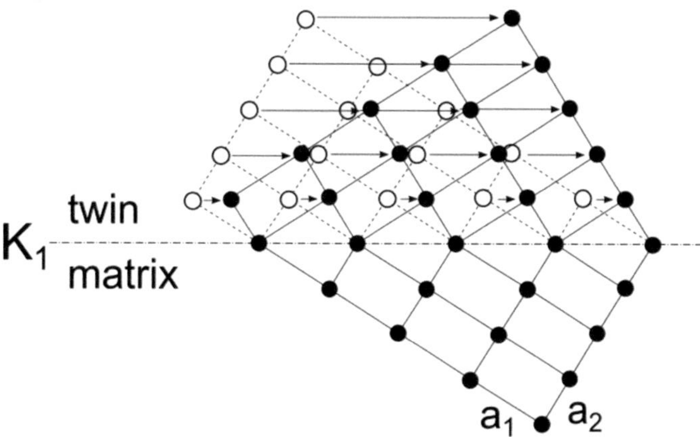


Figure 1.4: Textbook schematic of the twinning process. Atoms above the twinning plane K1 shear by a defined angle related to the ratio of a1/a2. The hollow spheres display the former positions of the atoms in the untwinned matrix [15].

When a crystal is free of dislocations, as it is the case in defect-free grown structures as nanowires, both ODP and DT have to start deformation with the nucleation of the first leading partial dislocation. The following step of either nucleating a trailing partial dislocation to obtain a full dislocation or the nucleation of another leading partial dislocation on an atomically adjacent slip plane, resulting in a two-layer microtwin determines the mode of deformation. The size dependence of partial dislocation activity has been modeled with respect to the ratio of stable to unstable stacking fault energies [35], and similarly the twin tendency has been described by the ratio of stable to unstable twin fault energy [36-38]. However, these criteria do not take into account the resolved shear stresses acting on leading and trailing partial dislocation [31]. Also, many studies only consider special loading conditions, e.g. at crack tips [37-39], without investigating the inherent material property. Although much work has been done a unified picture of the complex process of deformation twinning is still lacking. If compared to ODP in fcc crystals, much less is known about DT.

## Brittle materials

Materials which have large fractions of covalent or ionic bonds do not necessarily show plastic deformation upon mechanical stress RT. This is due to dislocations being relatively immobile in this temperature range and easy shear cannot occur. Instead, fracture takes place by breaking of atomic bonds [40], when stresses are large enough. The inherent brittleness reported for these materials is adverse in production and applications. Typically, fracture occurs at stresses well below the theoretical strength for breaking atomic bonds. This is owing to the fact that fracture in brittle materials usually starts at flaws, which are small discontinuities in the microstructure [41]. According to Griffith [42] crack growth is only promoted if the overall energy is reduced, balancing the driving force of the crack with the resistance of the material itself [40]. The strength of the overall structure is then determined by the largest flaw in the system. Since their distribution on the surface or in the volume is rather stochastic, the strength of a brittle material needs to be treated rather probabilistically than deterministically [43]. In these stochastic physical phenomena the probability of failure increases with the load amplitude and the size of the specimens [43], since a higher load will make a smaller flaw more critical and a larger volume will hold more chances to find a critical defect.

Weibull distributions are widely used to account for the probabilistic behavior and therefore to allow for designing of brittle materials. The Weibull theory [44] is based on a weakest link concept, where the specimen can be regarded as a chain with many links, each of them being a defect of different size. If the weakest link fails the entire structure fails. A further prerequisite for simple Weibull analysis is that critical flaws do not interact. An analytical form of the Weibull distribution is given in equation 5, where $m$ is the Weibull modulus describing the variability of the strength a component can withstand and $b$ is a function of the volume, surface area or some other characteristic parameter. The dependence of strength on specimen size is the most striking effect, since an increase in sample dimensions goes along with the extension of the chain, thus lowering the fracture strength. The fracture strength $\sigma$ of a structure can be determined by the knowledge of its volume $V$ or its surface area $A$ plus knowledge of the fracture strength $\sigma_0$ and volume $V_0$ or surface area $A_0$ of an additional specimen by the equations 6 below.

$$F = 1 - exp\left[-\left(\frac{\sigma}{b}\right)^m\right] \tag{5}$$

$$b = \sigma_0 \left(\frac{V_0}{V}\right)^{\frac{1}{m}} \; ; \; b = \sigma_0 \left(\frac{A_0}{A}\right)^{\frac{1}{m}} \tag{6}$$

## *Silicon*

At elevated temperatures inherently brittle materials can also deform by plasticity when dislocations become mobile. At high temperatures of 550 °C plasticity was found to occur in bulk Si [45]. The diamond cubic (dc) structure of Si is built by two interpenetrating fcc lattices. Glide of dislocations is strongly related to glide in fcc crystals with Burgers vectors of a/2<110> and {111} slip planes. However there are two types of {111} glide planes in this structure, a/12{111} narrowly spaced glide planes and a/4{111} widely spaced shuffle planes. In contrast to the former, the latter do not have stable stacking faults [46], whereas perfect dislocations can move on either type. Theoretical calculations predict slip on shuffle planes. In contrast, experiments conducted at high temperatures reveal partial dislocation activity pointing towards slip on glide planes [47] and so a fully understood picture is still not available.

## *Ionic crystals*

If a metal and a non-metal are combined, in many cases a complete transfer of one or more electrons occurs, leading to the formation of an ionic crystal. The energy barrier for transferring an electron from the metal to the non-metal is relatively low [48]. Ions are charged and attract each other. Many ionic structures form simple cubic lattices, where the most commonly known is rocksalt (NaCl). Since the work performed on ionic crystals in this thesis is restricted to sodium chloride, the fundamentals will be strictly confined to simple ionic structures of the rocksalt type. For information exceeding this section as well as information on different ionic structures the reader is referred to the comprehensive book of Sprackling [49].

If the radii of both atoms sufficiently differ ($r^+/r^-<0.723$) the rocksalt structure is obtained [16] with the lattice having the coordination number 6. $Na^+$ and $Cl^-$ ions form fcc lattices respectively, where each anion is surrounded by six cations and vice versa [50]. Thus the overall lattice consists of two interpenetrating fcc lattices of opposite ions [49].

Especially in the 1930s sodium chloride was subject to many investigations, which was mainly due to its availability and relatively simple structure. Therefore it was one of the first materials to reveal the disparity of theoretical and experimental observations of strength [51], as discussed earlier in this chapter for metallic systems. It is conventional wisdom to regard rocksalt and other simple ionic crystals to be inherently brittle at room temperature and to cleave upon the application of stresses, mainly due to our experiences mostly restricted to the daily handling of table salt. In 1924, Joffé was the first to show ductility of sodium chloride by immersing it into water (Joffé-effect) [52]. Later, Gorum [53] found the structure to be inherently ductile rather than brittle. Compression of NaCl in conjunction with x-ray analysis suggested plasticity by glide on {110} planes in <110> directions [54], which was confirmed later by Pratt [55] to be the primary glide system. A secondary glide system was discovered to be {100}<110>, where the CRSS to commence glide is about three times higher for NaCl, which as a consequence will only be initiated if {110} has a CRSS close to zero.

Cleavage and glide can be regarded as competing deformation mechanisms in the rocksalt structure. The glide planes only exist of cations or anions facing ions of the opposite sign (Figure 1.5) and glide exclusively happens along <110> directions, which are the most closely packed rows of equal ions in the lattice [49]. In contrast, cleavage is favored on planes that are most widely spaced and electrically neutral, e.g. {100} for sodium chloride (Figure 1.5). Whether cleavage or plastic deformation occurs is determined by the stresses acting on the respective planes. If CRSS is exceeding the critical normal stress, the crystal will deform by glide. Gilman *et al.* [56] showed that during cleavage the crack does not extend at a constant velocity and in regions of small or zero velocity plasticity occurs because stresses at crack tips can rise to levels sufficient to generate dislocations [56-58]. It should further be noted that the mode of mechanical response was found to be dependent on various parameters. A brittle-to-ductile transition was observed with increasing temperature and also with variation of the strain rate. Cleavage strength is constant over a large temperature range, however yield strength decreases with increasing temperature. Ductility was also found to be strongly dependent on surface treatment [49], which can cause the removal of preexisting cracks or embrittling surface layers [53, 59].

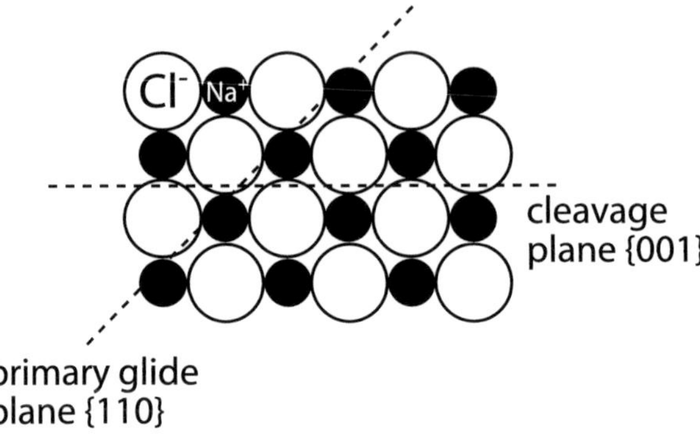


Figure 1.5: Sodium chloride lattice, showing primary glide and cleavage plane [49]. While glide planes only exist of cations or anions facing ions of the opposite sign and glide exclusively happens along <110>, cleavage is favored on planes that are most widely spaced and electrically neutral.

For glide along {110} <110> there are always at least two glide systems equally preferred. For uniaxial loading along [001] four equivalent glide systems exist for which one would expect multiple slip. Nevertheless one slip system predominated [60, 61] although slip lines from a second system could be observed in limited regions [61]. Hesse [60] reported that rocksalt crystals are softest for loading along [001] and hardening increases with approaching [011] or [111] respectively. Studies of cleaved NaCl compressed at RT revealed characteristic stress-strain curves that can be divided into different sections [55, 60]. Elastic loading is followed by stage I, which is characterized by a low and constant hardening rate. After a short transition region stage II is a region of linear hardening a factor of 2-3 greater than stage I. This is followed by a rate of decreasing hardening called stage III. In compression of sub μm crystals the hardening is absent up to 70% of deformation and the yield stress is increased by a factor of 15 [62].

Although rocksalt can deform by both cleavage and glide there is tremendous variation from specimen to specimen, depending on factors as different fabrication sources or preparation methods. This is further complicated by the affection of environmental conditions, such as humidity.

## 1.4 Small Scale Mechanical Testing

The mechanical testing of materials in small scales, where at least one dimension of the sample is in or below the µm regime, is markedly different from conventional mechanical testing procedures and is still lacking of standardization, e.g. no ASTM standards are currently available. Owing to the small dimensions and the accompanied difficulties in manipulation, imaging and measurement other approaches have to be taken in order to assure clean and interpretable testing. One of the major tasks is the application and measurement of very small quantities of load and displacement with proper resolution and the possibility to also reveal the smallest dynamic events that can occur during deformation. Furthermore, environmental isolation of the dedicated testing system is crucial, since already small oscillations can strongly influence the quality of the obtained results. The following paragraphs focus on some instrumented mechanical experiments and the related sample geometries that stimulated the progress in understanding the behavior of materials at these small scales.

Nanoindentation is a tool, which has gained much interest and is nowadays used for standard characterization down to extremely small volumes. Like in conventional indentation, a tip with known mechanical properties is pressed into a sample with unknown properties. Observations of the corresponding loading and unloading response are used to determine the elastic modulus $E$ and the hardness $H$. For these small indents, instrumented

Nanoindenters are usually equipped with a Berkovich (three-sided pyramid) tip (Figure 1.6A) and have to exhibit high resolution below a µN/nm [63]. Also the determination of the small contact area, which has to be laboriously identified by means of scanning electron microscopy (SEM) or atomic force microscopy (AFM), is circumvented by using a tip machined in high precision. For this known geometry the area can easily be calculated at each depth. From the slope of initial unloading the contact stiffness is calculated, which in turn is used to determine the elastic modulus of the tested material. Although this technique is applicable to a huge variety of materials, the localized and nonuniform stresses beneath the tip as well as the pile-up of material around the contact impression complicate the interpretation. Advances have been achieved by implementing continuous stiffness measurement (CSM) technique, where a small dynamic oscillation is imposed in order to provide continuous results and precise identification of the contact point [64]. Nanoindentation has been performed on all kinds of materials and has found extensive application in measuring thin films on substrates.

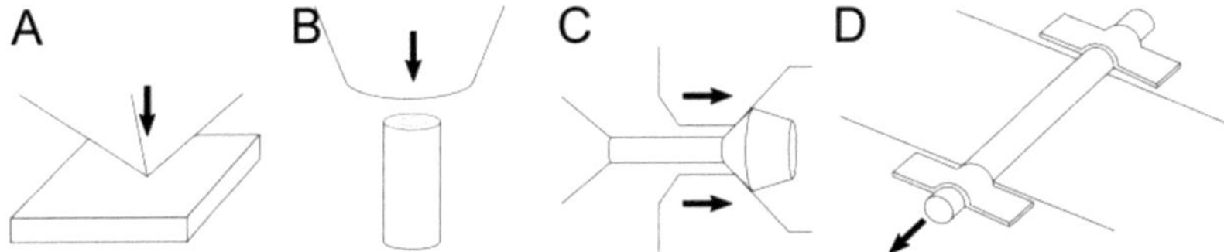


Figure 1.6: Schematic of various testing methods for small scale materials. Nanoindentation (A) employs the penetration of a precise tip into a surface. In micropillar compression (B) columns of various sizes are milled out of a surface and subsequently compressed with a flat tip. The direction of load can also be reversed by the application of a form locking grip and the compatible sample geometry (C). The testing of small structures inside an electron microscope almost exclusively is connected to gripping the specimen to a substrate by the deposition of a suitable precursor (D).

Thin films as building blocks for microelectronic and micro electro-mechanical systems (MEMS) have been subject to a variety of mechanical testing experiments over the last decades, borne by the need to understand their mechanical behavior in order to design reliable overall devices. The mechanical testing and its related advances and issues have been discussed amply in several review articles [65-68], which should be consulted for further information. Additionally, books of Ohring [69] and Freund and Suresh [70] cover mechanical properties and testing of thin films. In the following paragraphs only the most relevant aspects of this area will be mentioned. For bulge testing, which was originally developed by Beams [71] in 1959, a membrane or film is clamped to a nozzle and subsequently a uniform pressure is applied by using a fluid on one side of the film, causing the film to bulge in a biaxial stress-strain state [72]. The height of the deflection is then measured optically with a microscope or interferometer as a function of the applied pressure and facilitates the acquisition of stress-strain curves and residual stresses. Calculation of the deflection can be difficult at large states of deformation due to nonlinear behavior of the membrane [72] leading to strain distributions which cannot be explicitly assigned. Furthermore, sample preparation is important since the results are susceptible to small changes. Earlier work comprised samples that were prepared by stripping a film from a substrate for subsequent clamping or gluing [66], often causing irreproducible boundary conditions and initial stresses [67]. Nowadays this can be overcome by deposition of thin films and selectively etching away the substrate to obtain thoroughly framed films in reproducible conditions [67].

Substrate curvature measurements are another way to mechanically characterize thin films. Due to differences in thermal expansion coefficients, a curvature is induced by (i) residual stresses that can be directly measured or (ii) can be applied upon cooling and heating. If the thickness of the substrate is significantly larger than the thickness of the film the latter has to accommodate the mismatch entirely. The curvature can be detected by deflecting a laser beam. By the angle of reflection the radius of curvature of the substrate, which is proportional to the stress in the film, can be obtained. The thickness of the film is the only parameter of the sample to be known for the calculation of stress [73]. However, this method also has drawbacks, e.g. the temperature and strain cannot be varied independently and only small amounts of strain can be generated with this method. Thermally activated deformation mechanisms and the change from tension to compression between heating and cooling generate complex dependencies, which require thorough interpretation.

With tensile testing the most important mechanical parameters, e.g. yield strength and Young's modulus, can be directly extracted. Although large strains can be generated at isothermal conditions, this technique was applied rather rarely to thin films in the past decades, due to the difficulties in handling the fragile specimens. Their extreme sensitivity to flaws, owing to the high surface to volume ratio, made it difficult to maintain the integrity during removal from the substrate. If this can be achieved properly the measurements are performed free from the influence of the substrate. Another crucial parameter is the uniaxial alignment which can be hard to assure by manually handling these small specimens, but is key to obtain clean and interpretable data. Also very small quantities of forces have to be detected and strain measurement by conventional methods as strain gauges is impossible. To circumvent this Ruud *et al.* applied a thin grating to thin films to directly measure strain from the displacements of laser spots [74]. With the advent of micromachining techniques it was possible to produce damage-free thin films which are attached to a Si frame to maintain alignment and to allow for efficient and safe handling [75]. Films were mostly created by combining standard deposition technology with photolithography and substrate etching. After mounting the frame to an instrumented testing apparatus, the parts parallel to the tensile direction were cut while maintaining alignment [75]. Haque and Saif developed a system for tensile testing inside the SEM and TEM (transmission electron microscope) [76], where one side of the frame is concurrently used as the force sensing unit. Made out of single crystalline Si, the deflection of this part can be measured and the generated force can be calculated by the knowledge of its accurate dimensions. Furthermore, tensile testing of thin films has been conducted on compliant substrates, mostly of polymeric [77] or metallic [78] nature. In this case the sample handling is not critical and the sample can be strained to large amounts by conventional methods without strain localizations causing fracture [78]. Direct output of the mechanical response of only the film is not possible but can be achieved by measuring specimens with and without film in order to back out the force necessary to deform the substrate [79]. Schadler and Noyan [80] performed *in situ* XRD tensile measurements on Cu thin films deposited on compliant polyimide on a Ni substrate. At each strain increment they determined the stress from the change in lattice spacings.

With every step in technology new methods for characterization and testing emerged. The field of micropillar compression came along with the commercial availability of focused ion beam (FIB) microscopes. Uchic *et al.* [2, 81] were the first to report the usage of FIB milling to machine cylindrical compression samples out of bulk material. In this method the $Ga^+$ ions are used to remove material leaving a freestanding pillar. Subsequently the sample is compressed using commercial nanoindentation systems equipped with a flattened tip (Figure 1.6B). No handling of the specimens is needed since they simply remain an integral part of the substrate [2]. Samples can be serially produced in comparatively high throughput out of basically any bulk inorganic material [2] in various geometries. The location and the size of the specimens can be selected, which facilitates to shape single crystals and to probe the mechanical response of certain crystallographic orientations in the absence of grain boundaries. Also, it is straightforward to calculate engineering stresses and strains from the measured forces and displacements. However there are several weak points that impede the interpretation in micropillar compression. If the sample is milled from the top, the pillar is slightly tapered leading to non-uniform cross sections and hence to inhomogeneous stress concentrations and deformation [11]. By using the "lathe" [81] technique uniform cross sections can be obtained. However the influence of elastic deformation in the underlying substrate can falsify data in both cases. Additionally, misalignment and friction between indenter tip and specimen surface can result in erroneous data due to sample buckling and stress localizations. In consequence quantitative data should be checked critically. For clean testing the minimum sizes of pillars are limited to approximately 200 nm with aspect ratios between 2:1 and 4:1 [2] in order to prevent buckling. A major concern is the possible effect of surface damage by the penetration of $Ga^+$ ions, which will vary for different materials. The studies undertaken to elucidate this fact [82-84] show strong controversy and will be further revisited in chapter 1.5. For certain materials, micropillars also can be generated by chemically removing the matrix of a directional solidified eutectic [85], circumventing the issues that arise with ion beam milling. While all these compressions were accomplished *ex situ*, there is a trend to compress pillars *in situ* to concurrently observe the mechanical response [86].

Tensile testing avoids various constraints and limitations if compared to micropillar compression testing. It is the standard testing technique for bulk materials, integrated in American Society for Testing and Materials (ASTM) and European Norm (EN) standards. Tension, if performed uniaxially, provides a well-defined and nominally uniform stress and strain state that facilitates straight forward acquisition and interpretation of material parameters like yield strength and Young's modulus. Furthermore, tension is not affected by elastic instabilities, which allows the use of high aspect-ratio samples, like quasi-1D structures. In the 1950s, S.S. Brenner reported on an instrumented tensile testing setup for pulling metallic filaments with diameters in the micrometer regime [87]. In this setup he was able to place wires as small as 1 µm with good axial alignment and apply tension through a solenoid. While imaging with an optical microscope he was able to obtain qualitative stress-strain data. A year later, he presented a modified version of the first setup [88]. The key change was the implementation of a micrometer brake which allowed to stop the plastic flow right after yielding and therefore to prevent the specimens from fracture and enabling the observation of plastic flow at lower rates. Over the following decades there have been similar approaches to tension small fibers that are still visible in optical microscopes [89, 90].

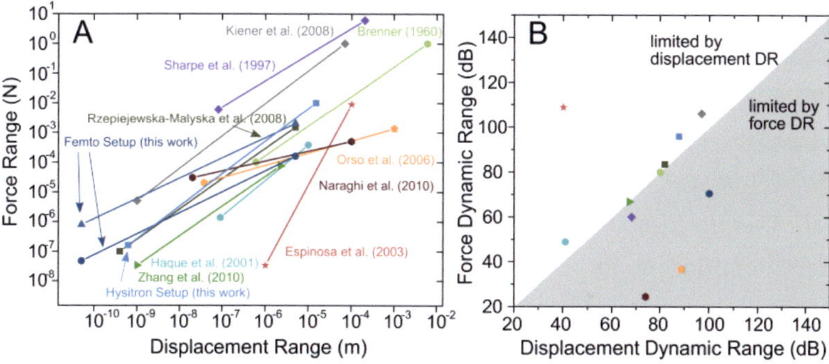


Figure 1.7: Comparison of the dynamic range of small scale testing instruments. In (A) the extreme values for load and displacement are shown for various testing setups [88, 91-98]. (B) illustrates the force and dynamic ranges of the testing techniques corresponding to (A), where the dynamic range is calculated by the standard definition DR = 20 x log(maxvalue/noisefloor) and plotted in dB.

Several platforms have been developed for electron microscopes to address the manipulation and tensile testing of structures that are too small to be imaged by conventional light microscopy. Actuation and measurement can be applied with various techniques that imply different limitations. However many of the conventional systems provide insufficient resolution of displacement and force and therefore new approaches have to be taken. Among the most used techniques for actuation are piezoelectric, thermal, electrostatic and electromagnetic systems. Yu *et al.* [99, 100] reported on a system which is combined out of several commercially available stages and can be used for both, manipulation and mechanical loading. Two opposing AFM cantilevers are employed for picking up individual specimens. Once placed in between displacement is applied using a piezo actuator. Simultaneously the cantilevers act as force sensors by imaging their deflection. Orso *et al.* [95] combined a piezoresistive AFM tip with a micromanipulator. Using FIB technology they were able to harvest, grip and tension individual beetle setae (Figure 1.6D). A major drawback for AFM based systems is the non-uniaxial motion of the cantilever, complicating the stress state and causing specimen bending at the grips with increasing displacement. MEMS fabrication techniques enabled the design and production of electrostatic [98, 101] and thermal actuators [101]. The compact chip design even makes these systems sufficiently small to fit into TEM holders [76]. Thermal actuation is achieved by resistively heating freestanding beams resulting in thermal expansion. High forces but only small displacements can be generated [94]. Major concerns comprise the heat conduction to the sample and non-uniaxial motion caused by thermal expansion in all directions [98]. If a voltage is applied between two sets of combs the electrostatic attraction causes displacement. By employing digital image correlation (DIC) high resolution measurements can be obtained for all chip based devices. Comb drives facilitate larger displacements but only relatively small output forces. Naraghi *et al.* [94] developed a "nanotractor" that is driven electrostatically by sequencing the actuation voltages based on the principle of an inchworm to tolerate larger displacements. In 2008, Kiener *et al.* [93] used FIB micromachining to prepare dogbone-shaped tensile testing specimens. A form locking grip was produced by ion beam milling and attached to the head of a commercial indenter (Figure

1.6C) for tensile testing. Like in microcompression samples the damage produced by ion milling and its effect on materials properties still remains unclear.

A good way of comparison of available testing setups is the evaluation of their dynamic ranges in force and displacement, describing the distance between the maximum allowable value and the noisefloor of the system, which is limiting resolution. Figure 1.7 is adopted and modified from [86] and shows the dynamic ranges of various setups in force and displacement (Figure 1.7A) and in dB (Figure 1.7B). Though in many cases the dynamic range of both parameters is almost equal, some setups are largely limited by the range of one parameter. However, if smaller size scales are to be investigated the key parameter is the resolution and noise-floor of the system limiting the minimum measurable quantity. For many of the above mentioned setups particular attention has to be paid to the alignment of the tensile specimens in the focal plane. Since from electron microscopy only 2D information is obtainable further efforts are required to assure alignment. For an imaging mode with low depth of focus thorough focusing can already help to reveal misalignment. In other cases tilting of the sample stage or a quick glance with the ion beam, which is mounted inclined to the electron beam can shed light on the orientation; however it should be kept in mind that pre-damage can be induced. Once proper alignment is guaranteed, tensile testing is a suitable method to obtain clean and versatile information on materials properties at the nanoscale.

## 1.5 Mechanical Size Effect

### *Metals*

A size effect describes a certain change in the response of a material caused by a reduction of internal structural length scale of bulk materials or the overall physical dimensions of a sample [2]. At bulk length scales, constitutive models are established, which are able to accurately predict the behavior of a material. Typically, the mechanical properties in these classical continuum models are independent of size. When the dimensions of a material are sufficiently small, say in the lower micrometer regime, the strength significantly increases. Conventional plasticity models are no longer valid and new models need to be created in order to accurately explain the mechanical behavior. The effect of size on the mechanical response can generally be divided into intrinsic and extrinsic constraints [102]. For the former the physical mechanisms are evoked by changes in microstructure whereas for the latter dimensional constraints are responsible. As already denoted in chapter 1.4, a huge variety of experiments has been performed on a large number of materials. Several models were proposed to account for the increase in strength, but a generally accepted picture is still missing.

There are several comprehensive review articles dealing with the most important findings and the understanding of mechanical size effects [11, 103-105] to which the reader is referred to for a more complete treatment. The results and mechanisms discussed in this chapter are focused towards plasticity in confined single crystals and therefore on the extrinsic constraints. However, it is still worth to start off with an intrinsic constraint, which was one of the first effects of structural size on the strength of a material. In the 1950s Hall [106] and Petch [107] reported on the impact of grain size on the yield strength. The dependence is given by a power-law relationship,

$$\sigma_y = \sigma_0 + kd^n \tag{7}$$

where $\sigma_y$ is the yield strength and $\sigma_0$ is the yield strength of the single crystalline material, $k$ is a constant, $d$ is the grain size of the investigated material and $n$ is a power-law coefficient of -0.5 in the case of the Hall-Petch relation. Similar power-law dependencies are generally used to describe the size effects. The increase in strength was first attributed to the pile-up of dislocations at grain boundaries. Today additional distinct models have been proposed, however the physical mechanism still remains unclear [105].

As one of the first in the research of small scale single crystals, Brenner reported on the tensile strength of single crystalline filaments, termed whiskers [87, 88]. Although Taylor [108] was the first to draw metallic filaments and to show their unique properties, Brenner was the first to release quantitative results on instrumented tensile testing of metallic whiskers. These single crystalline specimens were able to sustain stresses and strains that were never achieved in bulk metals and were approaching the theoretical strength of the material. Although there was considerable scatter, the strength was reported to increase with decreasing dimensions and to be inversely proportional to the diameter. The whiskers exhibited sharp yield points followed by extensive glide regions [88] that were significantly lower than the initial yield stresses. Since the whiskers were of high crystalline quality, large stresses had to be applied to cause the nucleation of the first dislocation. If fractured portions of the wires were restressed after the first test they even showed higher strengths fortifying a weakest-link behavior were a larger sample exhibits a larger probability of finding a critical defect. This was further enforced by the dependence of strength on the length of the specimens and the influence of statistically distributed surface defects [87] where dislocation nucleation becomes easier [87]. Recently, instrumented tensile testing was performed on Cu nanowhiskers inside an SEM [109]. These smaller specimens sustained stresses even higher than reported by Brenner and exhibited apparent brittle failure, which is not expected for this material.

Gruber *et al.* used Au thin films to explore the deformation behavior under one dimensional confinement [110]. They were able to obtain quantitative stress-strain data and to concurrently map the initial and evolving microstructure. Tensile tests on films with thicknesses down to 31 nm showed similar behavior, but differences in yield strength. For films thicker than 60 nm a strong dependence with a power-law exponent of about -0.53 was found. The microstructure showed deformation twins, which were decreasing in size with increasing film thickness up to 160 nm. Above this value almost no twins were observed. Although there was some experimental uncertainty a clear trend points towards a change to deformation twinning as the dominant deformation mechanism with decreasing film thickness.

Size effects in quasi 1-dimensional single crystals recently experienced a renaissance with the advent of micropillar compression testing. Several groups reported on an increase in flow stress with decreasing pillar diameter [2, 111-114], which is illustrated in Figure 1.8 for various face-centered cubic (fcc) materials that can be combined into a single band by normalizing flow stress $\sigma$ with the shear modulus $\mu$ [11]. For comparison the data of Brenner for Cu whiskers is included. All investigated fcc materials were found to roughly scale with $n$ = -0.6. This is valid from several tens of micrometers down to several hundreds of nanometers for single slip as well as multiple slip orientations. Despite the simple scaling in strength with decreasing sample diameter, a change from mechanical response known from bulk materials to a more stochastic yield strength and strain hardening response are observed. Stress-strain curves exhibit a serrated profile where elastic loading is intermittently followed by strain-bursts [115]. The intervals and the magnitude of single events is random suggesting the importance of the nature of single dislocations at these scales.

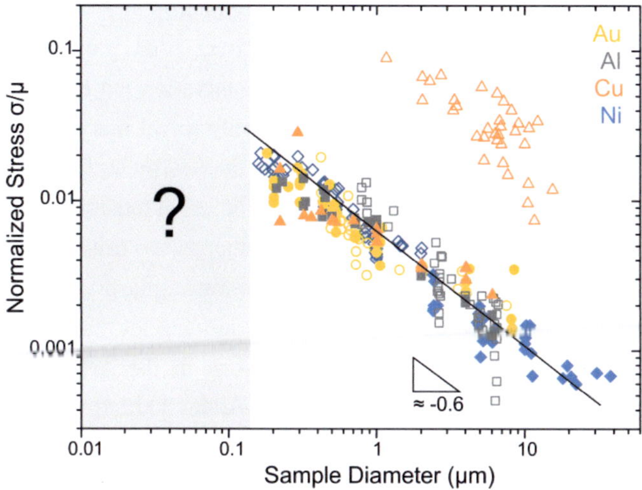


Figure 1.8: Normalized flow stress data for several micropillar compression experiments of fcc single crystals obtained from [11]. It can be seen that the sample size for all experiments roughly scales with a power-law exponent of -0.6. Additionally the Cu data of Brenner [87] is included. Due to experimental constraints the micropillar compression technique is not suitable to elucidate the size dependence of smaller size regimes (grey area).

Several attempts have been made to explain this size-dependent behavior, driven by diverse observations in the mechanical response and in the microstructure. These observations are further complicated since no standard evaluation technique exists. For example, different groups measure flow stress at different stages (2, 3, 5 or even 10%) of plastic flow. Further the interpretation of post-mortem TEM investigation can be affected by differences in foil plane of microcrystals [11] or irradiation damage, which will be picked up again later. In the following paragraphs the main findings will be summarized.

Conventional dislocation plasticity theory involves the interaction and multiplication of dislocations when stresses are applied to a material. However, if the dimensions of the sample are in the micrometer range, interactions of dislocations may be less important. Greer *et al.* proposed the increase in strength to be caused by dislocation starvation [112]. If the sample is small enough, mobile dislocations can only travel much shorter distances and predominantly leave the sample at the nearby surfaces, assisted by the attractive effect of image forces [114], before interacting with other dislocations. The replication of dislocations is suppressed causing a poverty of mobile dislocations in the sample. To continue plastic deformation, much higher stresses need to be generated for moving less favorable or to nucleate new dislocations. Residual dislocations in the specimens were attributed to be located on unfavorable slip systems [112, 113]. This mechanism was further encouraged by the findings of Shan *et al.* where, during *in situ* TEM investigations, they were able to obtain a significant decrease in dislocation density for pillars that had a high density of initial defects by solely applying mechanical stress; they termed this effect "mechanical annealing" [83]. Small pillars were left essentially dislocation free whereas larger samples retained little dislocations. Native oxide layers or FIB-induced damage did not seem to considerably affect the mechanical behavior.

A phenomenon, which in this context was first observed in the simulations performed by Rao and colleagues is the truncation of conventional double-ended Frank-Read sources (FRS) which results in single-arm sources [116-118]. Double-ended sources interact with free surfaces and truncated single arms of dislocations are created [116], which were found to be weaker than the corresponding FRS. The stress required to move these dislocations is related to the shortest distance from the pinning point to the free surface. If the sample size is decreasing, the spacing of possible pinning points for dislocation sources is decreasing, resulting in higher stresses required to move a dislocation through the lattice. The length of the largest single-arm source on a favored slip plane therefore determines the flow stress needed to plastically deform the crystal. The stochastic occurrence of dislocations as strain carriers in these confined volumes induces the observed variations in yield strength [117]. Further, the exhaustion and reactivation of single-arm sources, termed "exhaustion hardening" [118] is affecting the mechanical response in a stochastic manner. Although the simulations were performed at conditions far from experimental reality, Legros *et al.* were able to confirm the existence and to show the stable operation of single-arm sources in Al alloy microfibers under tension [119].

Kiener *et al.* performed similar tensile experiments on FIB-produced tensile specimens. Altering the width did not show a huge size effect, however the variation of aspect ratio did [93]. High aspect ratios of about 5:1 exhibited relatively low yield strength with negligible hardening and only a slight dependence on size [93], whereas short samples with aspect ratios of 1:1 showed stronger hardening and a size effect equal to micropillar compression. It was concluded that in long samples the successive operation and exhaustion of individual dislocation sources governs plasticity [93]. The strong size effect in short samples was related to constrained glide and the resulting pile-ups of dislocations due to reduced stresses at the transitions of the sample gauge sections. In contrast to previous findings the irradiation damage of $Ga^+$ ions was also suspected to cause a pile-up of dislocations [120].

Legros and colleagues [119] for fcc and Bei and colleagues [121] for bcc (body-centered cubic) showed that the amount of initial dislocation density does severely alter the mechanical response of the material. The dislocation arrangement of Al fibers was investigated before and after mechanical testing. It could be shown that the size effect was directly related to the defect density and the distance between the activated source and the surface, resulting in a transition from intermittent, pillar-like behavior, to the whisker-like behavior without strong plastic deformation. This accompanies the investigations by predeformation of pillars etched from a solidified eutectic alloy [85]. It was possible to show a transition from bulk like behavior for high prestrains (11%) over intermittent stochastic behavior (4%) to strengths approaching the theoretical limit of the material (0%). In contrast to other pillar experiments these results were independent of the width of the specimens. The difference in sample preparation also made it possible to investigate the influence of FIB irradiation damage on the mechanical response of the material. Contrary to the results reported above, the irradiation was significantly weakening the material, indicating that FIB milling introduces defects strongly reducing the strength of a material [84].

None of the models proposed above is proven and can be regarded as solely valid. It is obvious that different mechanisms are operating for different boundary conditions. Differences arise from the distinct starting configurations of the samples. Whereas whiskers are initially defect-free and require the nucleation of dislocations the micropillars contain initial dislocations that lead to an altered mechanical response. In many cases the various mechanisms may co-exist and therefore interact. It is further likely that for different size regimes or conditions the dominant mechanism changes.

### *Brittle materials*

Compared to ductile metals, other mechanisms have to be responsible for the size dependent behavior that is observed for brittle materials as ceramics, semiconductors or glasses. Since fracture is based on breaking atomic bonds, stresses are strongly affected by the size of pre-existing cracks. The first to show size effects in brittle materials was Griffith in the 1920's [42]. He reported on an increase in failure stress of glass tubes if the crack sizes are reduced and postulated that a crack will only grow if the crack extension will reduce the total energy (critical crack size). With the advent of micro electro-mechanical systems (MEMS) a significant part of research was focused towards Silicon. A comprehensive overview on poly- and single crystalline Si can be found in [122]. In the 1950's Eisner [123], Pearson [124] and Cook [125] reported an increase in tensile strength of Si whiskers with decreasing sample size. Hu tested Si wafers of different orientations and reported significant scatter in the strength data [126]. Bending tests on Si from nanometer to millimeter range [127] revealed a huge increase in fracture strength if the sample dimensions are reduced, while the elastic properties of the material did not change. Also a reduction in scatter with smaller dimensions was observed. All studies highlight the stochastic nature of fracture strength and some also state a reduction in variation with decreasing sample sizes. The findings of stochastic fracture strength can be explained by weakest-link statistics, which imply that the largest defect in the specimen is determining fracture strength. With decreasing sample size the probability of finding a critical defect is reduced and hence the strength increases. For small enough specimens the strength will therefore approach the theoretical strength of the material and, if the defect population is size dependent, concurrently result in a reduction of the scatter.

The advancements of experimental capabilities and computational power have led to new investigations on the mechanics of nanowires. However experiments and simulations show significant discrepancies in their results [128]. Experimentally, the Young's modulus was measured to increase below 200 nm [129] or to decrease below 25 nm [130]. In contrast a change in elastic properties is predicted to start below 10 nm for density-functional theory (DFT) and molecular dynamic (MD) simulations [131, 132]. Also a brittle-to-ductile (BTD) transition at room temperature is obtained in simulations for diameters smaller than 4 nm [133]. A similar effect has been observed experimentally at diameters below 60 nm in tensile experiments inside the TEM by dislocation-initiated amorphization

[134]. Östlund and colleagues [12] report on BTD transition in Si nanopillars at RT with a critical diameter of 310-400 nm. One proposed model comprised the size dependence that determines the deformation rate, which has to be sufficiently high for plastic deformation, whereas the other explanation was the nucleation of dislocations in the shuffle planes of the Si crystal. Apparently, significant differences exist between simulations and experiments but also within both, deriving from different interatomic potentials and parameters in the simulations or from fundamental differences of experiments.

A size effect was also reported for the mechanical interrogation of NaCl microwhiskers by Gyulai in 1954 [135]. He was able to grow ionic filaments with lengths up to 3 cm and subsequently test them by gluing single filaments onto a glass rod and measuring the deflection of the same using a microscope. With this information and the width of the crystals he was able calculate the fracture stresses and to observe a strong increase in fracture strength with decreasing diameter. Also in compression testing of NaCl crystals an increasing strength was reported if the sample dimension were reduced [51].

## 1.6 Nanowire Synthesis

The class of nanowires describes tiny elongated structures having diameters up to a few hundreds of nanometers. Since their lateral dimensions are much smaller than their length they are considered quasi 1D structures. Nanowhiskers are a special category of nanowires. Their growth by stacking individual atoms leads to the formation of faceted single crystals with a very high crystal purity and perfection. In the course of this thesis the term "nanowire" is used for simplicity.

The synthesis of nanowires is an individual discipline in the field of nanotechnology. For fabrication of nanowires there are two distinct approaches: top-down and bottom-up. The former incorporates the formation of small features by structuring a bulk material, employing combinations of patterning, lithography, etching and deposition. The latter describes the growth of these small structures by well-defined chemical or physical synthesis [4] from an atomic level. The main difference can be explained by simply removing atoms in top-down whereas adding atoms in bottom-up. Many different synthesis routes have been explored over the last decades. Nanowires have been fabricated by template assisted growth, through electro deposition in porous alumina or patterned polymer foils, the vapor-liquid-solid (VLS) process, described later in this chapter, or chemical vapor deposition (CVD).

Metallic wires have been grown the by reduction of their halides [87] in the middle of the last century. A precious survey was recently published by the producer of the metallic and ionic nanowires used in this work [14]. The metallic and ionic nanowires tested in this work were synthesized via physical vapor deposition under molecular beam epitaxy settings [109]. Essential to this mechanism is the vaporization of the desired material and its condensation on an appropriate substrate. Also the substrates were partly carbon coated by magnetron sputtering [109], which is necessary to facilitate growth of nanowires by providing preferred sites for nanowire formation. Nucleation of nanowires was found to commence close to impurities or defects. Another prerequisite is the elevation of the substrate to higher temperatures ($\sim$0.65 $T_M$) in order to enhance surface diffusion, which is also promoted by areas of different surface energies [14]. Once a nucleus is formed and grows, condensation and diffusion contribute to further extension of the crystal, only contributing to length if the place of atom incorporation is either at the tip or at the root of the nanowire [14]. The nanowires grow axially and radially during deposition, if the deposition rate is kept constant. The precise growth mechanism and the incorporation site are still unrevealed and subject to further research activity. However, it is clear that impurities, defects, surface roughness and surface energy [109] play a crucial role for the nucleation of nanowires grown by this method. Using the same principles, the fabrication of metal nanowires with different crystal structures, as well as ionic salts, is possible.

The growth of nanowires on W (Au NW) and Si (NaCl NW) at the above described conditions contributes to the high crystalline quality also shown by the faceted morphology in Figure 1.9C. These filaments are single crystalline and free of flaws and defects if examined by TEM (Figure 1.9C and D, courtesy of Gunther Richter MPI-IS, Stuttgart, Germany). Dimensions of the Au nanowires range from 20-300 nm in diameter and an average length larger than 10 μm. The <110> orientation along the axis is observed to be dominant for all sizes. The average spacing between single nanowires is usually more than one micrometer and the wires grow off the substrate at angles smaller than 30° with respect to the surface normal (Figure 1.9A), which makes them perfectly suitable for harvesting and manipulation. The NaCl wires grow in <100> directions exhibiting surface facets created by {100}-type crystal planes. Their average length is about 7 μm, while the width is ranging from 20 to 500 nm, as shown in Figure 1.9B. While the cross section of the Au nanowires is constant along the whole wire, the cross-sectional area of the NaCl nanowires is increasing towards the substrate.

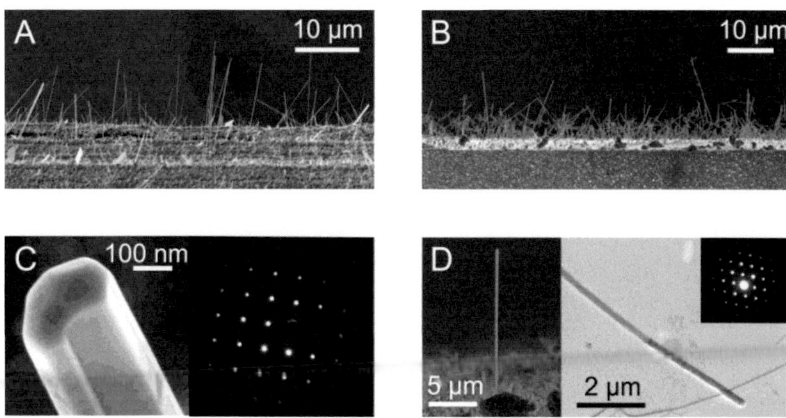

Figure 1.9: SEM micrographs of nanowire "forests" of Au (A) and NaCl (B) are shown. The high crystal perfection can be seen in the high resolution SEM and TEM diffraction pattern (C). A clean single NaCl wire is visible in (D). TEM micrographs courtesy of Gunther Richter (MP-IS, Stuttgart).

The silicon nanowires used in this work were synthesized by two different methods resulting in two different crystallographic orientations along their axes. The first set of wires was fabricated employing the vapor-liquid-solid process, first reported by Wagner and Ellis almost 50 years ago [136]. It reports the controlled crystallization of materials from vapor or liquid sources mediated by small seed particles of a foreign material. Since then, about a decade of thorough research was sufficient to reveal many of the fundamental aspects related to the VLS growth of Si wires [137]. Since the topic takes a renaissance due to the emergence of nanotechnology, Si nanowires are nowadays grown and investigated by many groups all over the world and there are several review articles addressing various growth methods and their fundamental aspects [4, 138, 139]. So, it is not the aim of the author to give a complete overview but to highlight the growth methods, which were used to obtain the nanowires tested in this work. A schematic of the VLS mechanism is shown in Figure 1.10B, for better understanding the binary phase diagram of Au and Si can be obtained from Figure 1.10A. As a first step Au nanoparticles are placed on a Si substrate (I) through e-beam deposition of a discontinuous Au film [13]. After the Au catalyst template is obtained it is brought to a metal-organic chemical vapor deposition (MOCVD) reactor. There it is heated above the eutectic temperature and a liquid Au-Si alloy droplet forms (II). This serves as a preferred site for the decomposition of the gaseous Si precursor which is pumped into the system. Gaseous semiconductor reactants can be created by various approaches. In most cases silane ($SiH_4$) is the vapor phase precursor of choice [140]; it will crack at the surface of the particles forming eutectic droplets (III). Continued supply of the reactant into the liquid droplet will cause supersaturation of the eutectic, leading to nucleation (IV) of the solid semiconductor [136]. Continuous supply of silane causes further precipitation through the liquid-solid interface, leading to further elongation of the nanowire (V) by continued Si incorporation into the lattice [141]. Through careful control of the growth parameters, diameter and electronic properties can be determined [142]. In Figure 1.10C a top-down SEM micrograph of a characteristic sample can be seen. The brighter droplets at the tips of the wires are the remaining Au catalyst, which is also shown in the inset. The crystallinity of the catalyst is visible through the faceted surface.

Diameters of the wires range between 80 and 500 nm with an average length of 10μm. From Figure 1.10D it is also visible that the carpet of Si wires grown by this method is very dense, complicating manipulation and isolation of individual nanowires. Further no strict vertical epitaxy on the (111) surface of the Si wafer is observed. The wires appear to grow in all possible <111> directions, one normal [111] and three <111> out of plane orientations with respect to the substrate [13]. Usually the wires exhibit an oxide layer of at least 1-3 nm [140], whereas Wu *et al.* report of oxide free wires, as well as a transition in growth direction to <110> for diameters smaller than 10 nm [143]. A second batch of Si <111> oriented nanowires was grown on silicon-on-insulator (SOI) wafers, according to the modified VLS process reported in [144]. This method produced longer and thinner nanowires exhibiting a cleaner surface without strong decoration.

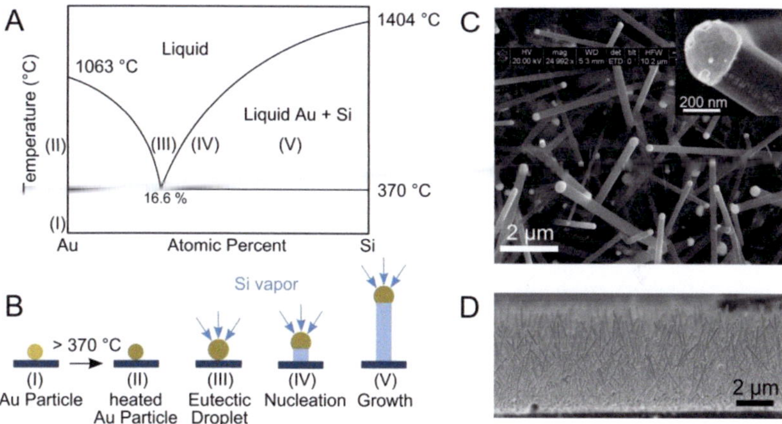


Figure 1.10: Si nanowire synthesis by the vapor-liquid-solid method. (A) and (B) show the VLS growth. Au particles are heated above the eutectic temperature of the binary Au-Si alloy, shown in (A). By supplying a gaseous precursor, which decomposes at the liquid droplet a eutectic is formed. Further supply causes the precipitation of Si and the axial growth of a nanowire. SEM micrographs show a top-down image (C) of <111> oriented Si nanowires. The inset shows a high resolution image of a single wire with the Au particle on top where the crystallinity can be seen by the facets on the surface. A side view of the carpet emphasizing the high density is shown in (D),

A different approach was employed in order to obtain Si nanowires with <100> axial orientation. Since these do not grow bottom-up because their orientation is not energetically favorable, they have to be created by a top-down process. In this method, mostly referred to as metal-assisted chemical etching [145], it is exploited that, by using wet chemistry, Si is selectively etched at the interface to a metal [146]. The fabrication route comprehending five distinct process steps is schematically illustrated in Figure 1.11A. First a monolayer of polystyrene beads is deposited on the Si surface (I), usually achieved by spin coating [147], where the beads form a close-packed hexagonal pattern. Subsequently, reactive ion etching (RIE) is performed to tailor the size of the polystyrene beads (II). The metal is then deposited on the modified surface, filling the interstitials of the polystyrene layer (III). After removal of the beads a continuous metal film with periodically distributed nanohole arrays is obtained (IV). A mixture of hydrofluoric acid, hydrogen peroxide and water [148] is then used to subsequently transfer the pattern to the Si underneath (V). Although the details of the catalytic phenomenon are not fully understood, it is generally accepted that the Si surface is passivated against dissolution unless a catalyst (metal) is available [149]. Since the etching is confined to the interface between Si and the metal, this process is a simple and effective way for the creation of vertically aligned nanowire arrays, which exhibit uniform crystallographic orientations, determined by the orientation of the wafer. Diameter and placement are highly controlled by the modification of the polystyrene beads [149], whereas the length is strongly dependent on the etch time. The etch rate decreases linearly with increasing aspect ratio of the resulting nanowires [147]. Figure 1.11B shows a SEM micrograph of a <100> Si nanowire carpet with diameters of 100 nm and lengths of 4 µm. In the inset the metal (Au) mask (bright layer at the root of the nanowires) can be seen at the bottom.

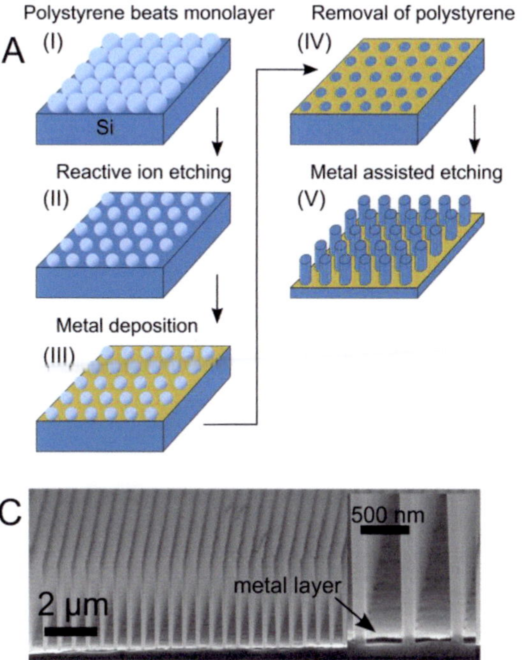


Figure 1.11: In (A) schematic of the metal-assisted etching process is shown. Polystyrene beats are used to pattern a Si surface with Au. The silicon is then exclusively etched at the Au-Si interface leading to arrays of nanowires. (B) shows a large array of <100> oriented nanowires produced by metal assisted etching. In the inset the metal mask can be seen at the root of the nanowires.

# 2 Experimental Techniques & Procedures

The following chapter focuses on techniques and procedures developed in order to perform *in situ* mechanical and electrochemical characterization, employing electron- or ion-beam microscopy. Dedicated devices needed for harvesting, manipulation and alignment as well as the application and measurement of load and displacement of specimens which can have diameters down to tens of nanometers are presented. The assembly of these devices resulting in two different setup configurations is shown and evaluated and the execution of a tensile test is explained step by step. The method of image-based strain measurement by DIC is highlighted.

## 2.1 Nanomechanical Testing

### *2.1.1 Components*

Since optical white light microscopy is only applicable to structures which have dimensions not significantly smaller than the minimum wave length of visible light, electrons or ions have to be used to facilitate accurate and interpretable testing of a nanowire. For the *in situ* measurements, a FEI Nova Nanolab 200 dual-beam scanning electron microscope and focused ion beam is used (in the following we refer to the entire system as FIB for brevity). The electron column is equipped with a field emission gun (FEG) which allows for high-resolution images of a sample surface with a nominal resolution of ~1 nm. Multiple detectors for secondary electrons (SE) and backscattered electrons (BSE) facilitate acquisition of as much information as possible. The ion gun emits $Ga^+$ ions and has a nominal resolution of ~8 nm. Owing to the different interaction and the higher mass of the $Ga^+$ ions compared to $e^-$ they can also be used for precise and highly localized removal of material. Further the dual-beam microscope is equipped with a Gas Injection System (GIS) to provide e-beam (EBID) or i-beam induced local deposition (IBID) of material. On the wall of the vacuum chamber various customized feedthrough connectors are installed, providing electrical connections for remote control of the various components and data acquisition.

## Manipulator and Positioners

For manipulation, transfer and alignment of nanostructures dedicated devices are needed which allow for additional degrees of freedom to the ones given by the microscope stage. Inevitable features comprise a very accurate motion in the range of a few nanometers to not damage the specimens but on the *other* hand a large overall range of motion is required to compensate the motion being done by the microscope stage. Furthermore, they have to be compatible to high-vacuum (HV) environment.

Suitable devices available on the market are piezo-driven inertial drive systems. The driving mechanism consists of a piezoelectric motor allowing for two different modes, coarse and fine movement. This mechanism of motion mainly consists of three parts: the piezo which is capable of executing expansion and contraction, a stator connected to the piezo and a slider clamped to the stator and held in place by friction. A schematic of this configuration is shown in Figure 2.1. By applying a voltage the piezo crystal expands or contracts and can be used in two different ways.

For coarse motion, further described in Figure 2.1, a sawtooth voltage profile is applied to the piezo. During the flat slope the piezo expands slowly (see Figure 2.1B, voltage sweep from 1 to 2) causing coherent motion of the stator and the slider due to frictional constraints. After having reached the final elongation, the piezo is rapidly contracted by a steep slope (see Figure 2.1C, voltage sweep from 2 to 3), where the inertial force exceeds static friction leading to a net displacement $\Delta x$ between the rigid frame and the slider. $\Delta x$ is always a fraction of the final deflection d, increasing with the mass of the slider. Then this sequence can be repeated, leading to a "ratchet"-type of motion, to a theoretically infinite distance, yet being limited by the geometrical constraints of the device. The step size and velocity can be determined by the applied voltage and the frequency. Although the resolution of this type of motion is ample for most microtechnological applications, this "slip-stick" motion is not fine enough to allow for exclusive manipulation of single nanostructures. Fast inertial movements, which could cause invisible damage to nanostructures, are not visible in the majority of cases due to the comparably much slower scan rate of the imaging device. To avoid the uncertainty of possible predamage, the devices can be operated in fine motion. Here the piezo is only scanned using relatively flat slopes, causing a deflection of the stator but leaving the slider unmoved relative to the stator. This allows for smooth control in the nanometer range and approach and contact without any uncertainty.

The two different devices used in this thesis are a Kleindiek MM3A Nanomanipulator and Attocube ANP Nanopositioners. Pictures of both can be seen in chapters 2.1.2 and 0. The former is used for manipulation of nanostructures and therefore is equipped with a fine Picoprobe tungsten needle at the distal end. The manipulator offers three degrees of freedom allowing for non-Cartesian movement; rotational movement in horizontal and vertical direction and linear movement along the main axis of the manipulator. For alignment of the sample to the force sensing device a stack of three Attocube Nanopositioners is used. Each of them features one linear degree of freedom, so that the combination of two horizontal (ANPx101) and one vertical (ANPz101) positioner results in movement along Cartesian coordinates. Their maximum travel range is 5 mm each in coarse mode and 5 μm in fine mode. By only using the fine motion, the horizontal positioners can also be used for the application of displacement in tensile testing (see chapter 2.2.3).

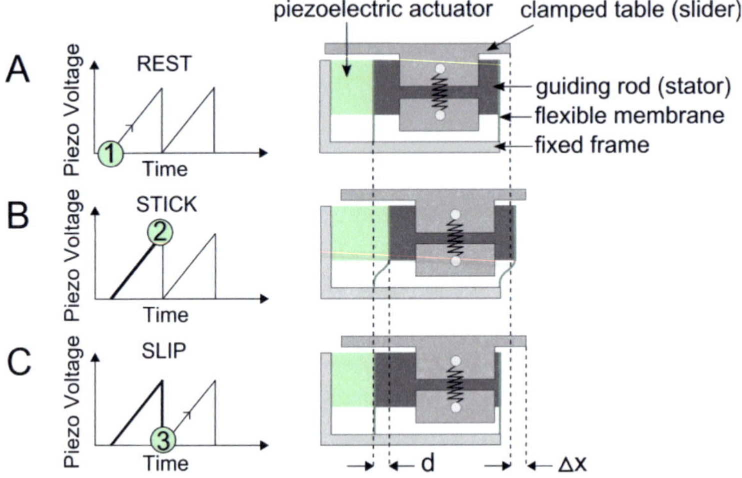


Figure 2.1: Schematic of inertial drive mechanism. The flat slope in (B) causes slow deflection of the piezo and coherent motion of the clamped table. By application of a steep slope the piezo deflects rapidly (C) overcoming the frictional force limit between the table and the guiding rod. This causes a net movement of Δx. By repeating this, large ranges can be covered with high resolution stepping.

## Hysitron Nanomechanical Transducer

One of the most important components of any mechanical testing system is the transducer. It can comprise actuation as well as measurement of force and displacement, performed by two separated or combined units. Since application and measurement in such small quantities is non-trivial, a huge emphasis has to be put on the design of hardware, software and the calibration. This transducer is capable of both, electrostatic force actuation and capacitive displacement sensing and was published first by Bhushan *et al.* [150]. The system used in this work is a design iteration of the one used by

Rzepiejewska-Malyska and colleagues [96]. It basically consists of a force actuation and a displacement sensing core, which is shown in a cross-sectional view in Figure 2.2A. These cores both are 3-plate capacitor units each having two fixed outer plates (drive plates) and a center plate which is suspended on springs (Figure 2.2B). They are fixed to a central rod covered by a ceramic sheath to provide electrical isolation [86]. Standard and customized probes can be threaded onto the distal end of the central rod. Conduction through the central rod is provided to minimize charging effects from electron beam imaging.

For electrostatic actuation a known DC voltage is applied between one of the drive plates and the center plate, moving the center plate towards the particular drive plate. Depending on which drive plate is used the transducer can be operated in tension or compression. Since the nominal gap between the center plate and either outer plate is relatively small (100 μm) and the plates are parallel, the electric field varies linearly over a certain range. The generated electrostatic force can be derived as follows: It depends on the charge $Q$ and the electric field $E$ in a linear fashion (see Equation 8). In turn the electric field can be expressed by the applied voltage $V$ and the distance $d$.

$$\vec{F} = \frac{1}{2} Q \cdot \vec{E} = \frac{1}{2} Q \cdot \frac{V}{d} \tag{8}$$

With the analytical term for the charge $Q$ (Equ. 9) and the capacitance $E$ (Equ. 10) where $\varepsilon_0$ is the electrical permittivity, $A$ the electrode area and $d$ is the electrode gap, the equation for the generated force can be derived in Equation 11.

$$Q = C \cdot V \tag{9}$$

$$C = \varepsilon\varepsilon_0 \frac{A}{d} \tag{10}$$

$$F = \frac{1}{2}\varepsilon\varepsilon_0 \frac{V^2}{d^2} \tag{11}$$

If $d_0$ is the nominal electrode gap (no voltage applied) and $\delta$ is the displacement from $d_0$ the equation can be rewritten to [86]:

$$F = \frac{\varepsilon_0 A}{2d_0^2}\frac{1}{(1 - \delta/d_0)^2}V^2 = \frac{\kappa_0}{(1 - \delta/d_0)^2}V^2 \tag{12}$$

$\kappa_0$ is the electrostatic force constant at the nominal gap. As can be seen in equation 11 the electrostatic force is proportional to the squared voltage and inversely proportional to the square of the capacitor plate gap. During actuation the gap between the electrodes will change and so does the electrostatic force.

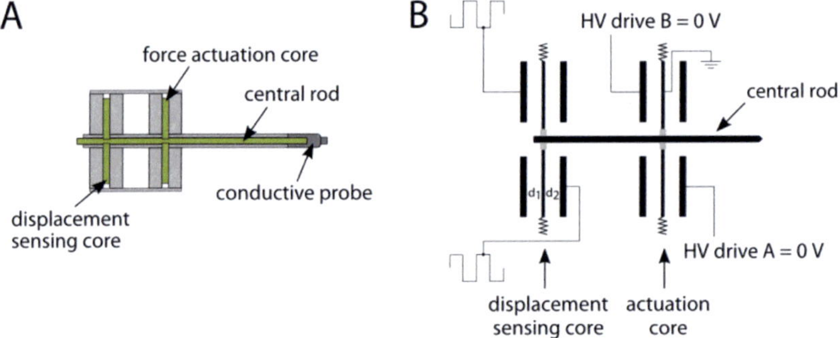


Figure 2.2: Schematic (A) of the transducer design showing the two different cores for actuation and displacement sensing. Both cores are connected to the central rod which carries the probe. In (B) the three-plate capacitive design can be seen.

For the measurement of displacement AC signals, where one signal is shifted 180° with respect to the other, are applied to the outer plates of the sensing core (Figure 2.2). They both have a frequency about 125 kHz, much higher than the mechanical bandwidth of about 100 Hz, to not actuate the transducer [86]. The signals are cancelled out by each other at the rest position $d_0$, however during actuation of the transducer the center plate of the displacement sensing core senses the electric potential at its position and the displacement from the rest position can be determined.

The maximum force that can be generated by the device is close to 10 mN and the maximum available displacement is 15 µm in either direction.

Calibration is done in two different steps. For determination of displacement the transducer is mounted in the operational configuration (horizontal) and the displacement for applied voltages is obtained using laser interferometry. This is done over a range of $\pm 15$ µm where the transducer is known to be linear [86]. The spring stiffness ($k \approx 360$ N/m) is determined by the suspension of deadweights from the transducer tip mounted in vertical direction. Changing the mass of the tip or having different environmental testing conditions changes the rest position of the center plate. Therefore the electrostatic force constant at zero volts $\kappa_0$ and the rest gap of the capacitor at zero volts $d_0$ are determined before testing. In this case the transducer is only actuated against the internal spring of the transducer without the tip being in contact to a sample.

## FemtoTools Force Sensor

The suitability of using capacitance based force measurement for resolving forces in the nN regime is state of the art. Production of these devices is relatively cheap and easy utilizing microfabrication technology which is a key technology in the manufacture of microchips and micro-electro-mechanical systems (MEMS).

The operating principle of the force sensor used in this work is based on capacitive deflection measurement, which is also the basis for the displacement sensing of the transducer shown before. A picture of the force sensing unit can be seen in Figure 2.3 [151]. The device consists of an arrangement of beams (comb) machined from single crystalline Silicon (Si). Both outer combs are stationary, but isolated from each other (Figure 2.3). Together with the movable combs in the middle, which are connected to the body by Si beams acting as springs, they form the capacitive unit, a differential three-plate comb drive. If a force is applied the distance between the gaps will change and so does the capacitance.

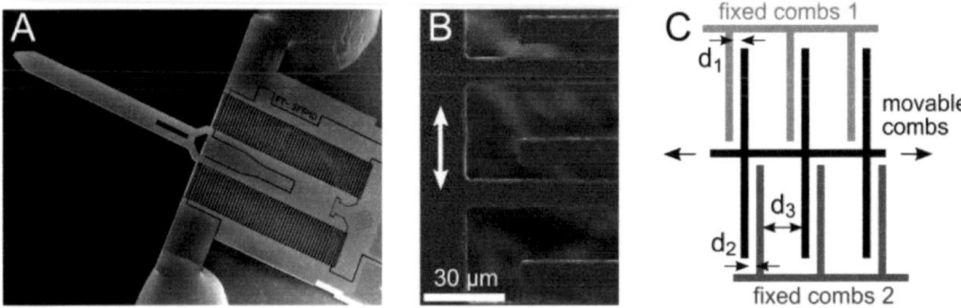


Figure 2.3: SEM micrographs of micromachined FT-S540 force sensing unit (A) [151] and combs made of single crystalline Si (B). (C) shows a schematic of the differential triplate comb drive which is used in this device. If a force is applied the movable part is deflected thus causing a change in capacity.

When an AC signal is applied to the outer capacitors, a voltage divider is formed (see Equation 13) [152]. The capacitor plates are spaced equally in a way that $d_0 = (d_1+d_2)/2$, the displacement of the middle plate is represented by $\Delta d$. Since the plate spacing is $d_1 = d_2 \ll d_3$, the additional parallel capacitance is minimized in order to maintain linearity. The proportionality of the output signal to the middle plate displacement is shown in Equation 14.

$$V_{out} = V_s \frac{C_1 - C_2}{C_1 + C_2} \qquad (13)$$

$$V_{out} = V_s \left( \frac{\Delta d}{d_0} \right) \tag{14}$$

The stiffness of the sensors can be determined by the dimensions of the springs being used. Modeling the force-deflection gives the results in Equation 15 [152] where $F$ is the total applied force, $E$ is the Young's modulus of the respective material and $l$, $w$ and $t$ (length, width and thickness) represent the dimensions of the spring.

$$\Delta d = \frac{F l^3}{4 E w^3 t} \tag{15}$$

The force sensors have a maximum displacement $\Delta d_{max}$ of 2 µm. For these relatively small displacements the linear relationship between the gap change and the change in capacitance is maintained. Each sensor is individually calibrated. Depending on the spring dimensions the sensitivity and the maximum applicable force can be tuned. The sensors used in this thesis are FT-S270 and FT-S540. The main differences are shown in table 2.1.

| Model | Maximum Load | Sensitivity (1000 Hz) | Stiffness | Res. Frequency |
|---|---|---|---|---|
| FT-S540 | 600 µN | 0.3 µN | 90 N/m | 2000 Hz |
| FT-S270 | 2000 µN | 2.0 µN | 1000 N/m | 6400 Hz |

Table 2.1: Specifications of FemtoTools force sensors as given by the manufacturer.

## 2.1.2 Transducer Setup

For *in-situ* testing of nanowires the whole setup needs to be assembled inside the vacuum chamber of the dual-beam microscope. Assembly is constrained by the limited amount of space and existing fixation points inside the chamber. Also, delicate parts of the microscope, which can easily be damaged, are situated in the vicinity of the setup. This and the uniqueness of the experiment require the design of a customized solution. A setup has been established prior to this work by Daniel Gianola at KIT. Details of this configuration and the evaluation of its performance are discussed in Appendix A.1. Since the rather compliant design makes the original setup highly susceptible to picking up mechanical noise, a new design was developed as part of this thesis work.

This new design primarily provides a strong improvement in stiffness between transducer and sample. Due to this increased stiffness, the noise rejection is improved since the sample and the transducer better stay in phase when the setup is subjected to external noise which is always the case to some extent. In this modification, the degrees of freedom and the range of motion need to be maintained. To avoid danger to delicate parts of the system and to ensure a convenient installation, the handling capabilities were also improved by the new design.

For the iteration of the existing setup, the rearrangement of the Attocube positioners and the transducer helped to eliminate the cantilever like design and therefore improved the stiffness. As can be seen in Figure 2.4A, the stack of positioners is divided and the transducer is brought to the translational movable part of the microscope stage, associated with the advantage of excluding the Kleindiek bridgemount, only clamped by the friction of screws, as the support. With respect to the stiffness the dovetail joint will no longer be used. Instead the transducer is mounted rigidly at the bottom of the housing below its center of mass, realizing the stiffest configuration possible.

For the new setup, a plate is machined that can be attached to the FIB translational stage by four screws. There is a hole in the physical dimension of the FIB rotational stage cut out to allow for free motion of the rotational part of the microscope stage and an elevation at one end where the z-Attocube is attached. On top of the z-Attocube there is another smaller plate where the transducer is gripped to by two clamps which fit tightly around the housing of the transducer. The xy-Attocube stack is mounted to the microscope rotational stage through a circular plate. On top of the stack, a newly designed sample holder is attached. It allows for mounting of conventional SEM-stubs in two configurations, perpendicular with respect to the e-beam for tensile testing as well as parallel to the e-beam to perform indentation and compression. Since there is significantly more space covered on the microscope translational stage, the immobile bridge-mount, to which the Kleindiek manipulator is attached, is shrunk to provide enough range of motion. All parts of the system are tightly held together by screws. In order to warrant compatibility to ultra high-vacuum (UHV) environment all parts are machined out of stainless-steel (1.4301) or an Aluminum-alloy (AlMg3).

A computer-aided-design (CAD) drawing in Figure 2.4B highlights the available degrees of freedom. In the new configuration, the transducer is only independently moveable in z direction, allowing the tip to be brought to the eucentric height, which is the tilting axis of the microscope stage and by design the coincident point of electron and ion beam. Starting at a working distance (WD) of about 8 mm it can be elevated up to 3 mm, which is also helpful to maneuver the tip to a secure position when other parts of the system need to be moved within large ranges. Since the transducer is now coherently moving with the microscope translational stage, tip and sample have to be approached by a coupled motion of xy-Attocube stack plus additional z and rotational motion of the microscope stage. The Kleindiek manipulator is still located in the same position as before.

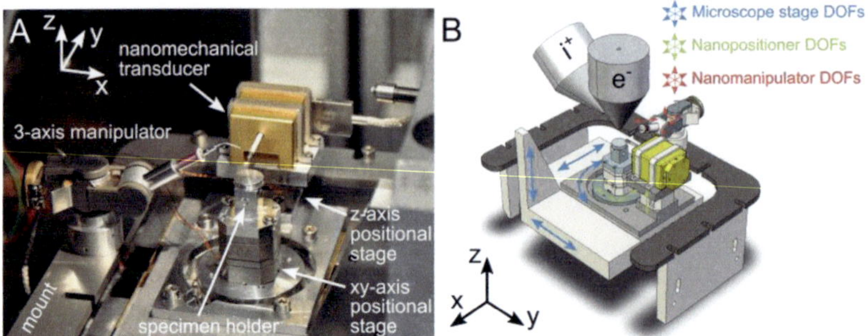


Figure 2.4: New design of Hysitron setup inside the vacuum chamber of the dual-beam microscope. (A) shows an optical image of the assembly with the labeling of all crucial parts. The stiffened configuration by clamping the transducer below its center of mass is visible. In a 3D-CAD drawing (B) the degrees of freedom for all moveable parts are highlighted.

Noise floor measurements have also been conducted for the new setup to elucidate the performance of the stiffened design. The system was tested in the same configurations as the previous setup, described in Appendix A.1, using the same parameters. The FFT spectrum in Figure 2.5 with the transducer mounted to the black holder is improved mainly because it was partly encapsulated from the environment. This is also shown in Figure A.1.4 where the noise floors for force and displacement are plotted. For comparison the FFT spectra of the first setup are visualized in grey (Figure 2.5). There it can be seen that the noise around the mechanical resonance peak is diminished considerably in air indicating a strong improvement in stiffness. This is also the case for the spectrum inside the vacuum chamber besides one distinct peak which will be discussed further in Appendix A.1.2.

The improved noise floors for the design iteration in comparison to the values obtained for the first setup can be seen in Figure A.1.4. The noise of the new setup on the air table is close to the transducer without setup placed in the black storage holder (cf. A.1.1), corroborating a huge improvement in stiffness, when compared to the previous assembly. Also the values in vacuum finally are amended, but never approached the values in air. The possible issues on the hardware side were already addressed by designing the stiffest configuration possible and by the built-in pneumatic isolation of the microscope from the ground.

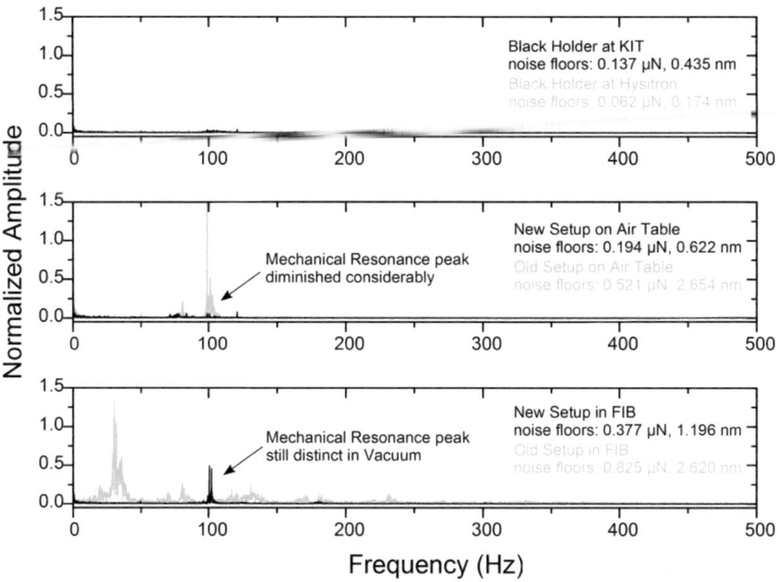


Figure 2.5: FFT spectra of various testing configurations for the new design. The noise existing in the first setup (grey curves) is diminished significantly indicating an improved stiffness of the system.

Since accurate measurements in nanomechanical testing can be tremendously dependent on the existent noise, the improvement of the results strongly hinges on the minimization of vibration in the system. There are many possible origins for noise, especially in vacuum environments, such as mechanical vibrations, electrical noise from the AC line source, exhibiting characteristic frequencies, and electrical noise from stray electromagnetic fields [86]. For eliminating these highly undesired effects there are basically two possible routes, inhibiting the origin of the noise or active compensation. For systems operated in vacuum this is particularly necessary since the natural damping of the ambient air is no longer existent. The issues being present if the system is brought into vacuum environment are shown in Figure 2.6. Once the damping of the air disappeared, oscillations around the resonance frequency of the device can increase the noise dramatically.

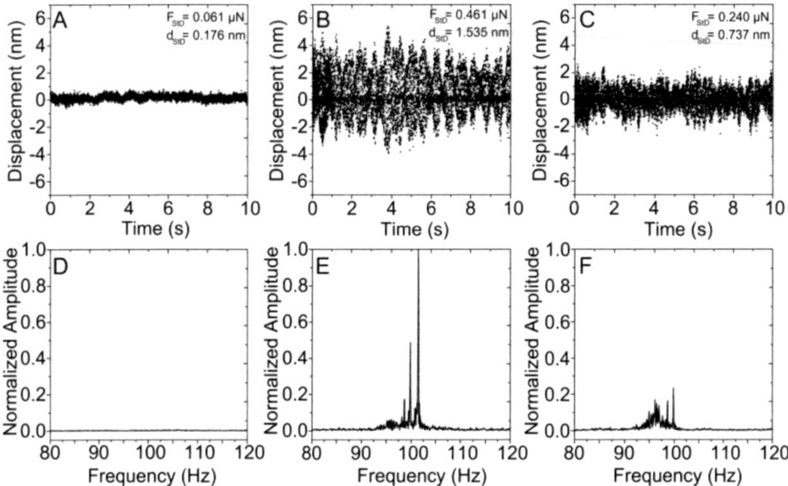


Figure 2.6: Noise characteristics of the transducer in air and in vacuum, showing different noisefloors for open-loop measurements in air (A), in vacuum (B) and using Q-Gain algorithm in vacuum (C). The corresponding FFT spectra (D, E, F) show significant differences in the amplitude of the natural resonance frequency around 100 Hz. By actively compensating the oscillations the noise can be reduced by a factor of 2.

Besides the already accomplished improvements, other possible counteraction to further reduce the noise comprises the implementation of suitable algorithms in hardware and software architectures. Figure 2.6 displays the different noise characteristics for open-loop mode in air and vacuum. The noise increases by a factor of eight due to oscillations around the mechanical resonance frequency of the device, as can be seen in the time domain (Figure 2.6B) and the corresponding FFT spectrum (Figure 2.6E). By exchanging the controller to a newly developed digital signal processor (DSP) controller and data acquisition unit operating at a fast internal feedback rate of 79 kHz as well as integration of a new control algorithm which addresses the quality factor of the system, the noisefloor could be further reduced (Figure 2.6C) and in the FFT plot in Figure 2.6F a strong reduction of the amplitude is visible.

### 2.1.3  Microforce Sensor Setup

For the investigation of size effects the aim is to access smaller regimes by keeping the ability to still uncover the mechanical response. This task is mainly limited by the performance (noise) of the equipment that is used to measure the behavior. In order to be able to go to smaller sample dimensions, by maintaining the accuracy, another setup was developed using the FemtoTools Force Sensor (see Chapter 2.1.1) as the load sensing unit. There are several advantages to accompany the utilization of this device. Besides the high accuracy, the device is only capable of force sensing and therefore much smaller in dimensions. Actuation is accomplished by fine motion of the Attocube positioner piezo and does not require sophisticated feedback control. By designing a very compact setup a good range of motion and a low noise level can be gained. Also the mounting into the vacuum chamber can be done more efficiently and furthermore, the purchase costs of these devices, including the data acquisition unit, are much lower than the Hysitron equipment.

The design basically consists of two towers, which are mounted to a plate that can be attached to the rotational part of the microscope stage, as shown in Figure 2.7. The tower on which the sensor is fixed is about the same dimension as the sensor. In order to have the tip and therefore the sample as the highest point of the setup, the sensor is mounted upside down. On top of the tower the topography of the Femto Tools Sensor and a rail that guides the sensor are milled to provide rigid alignment by a form locking connection. The sensor can be slid in and fixed by a single screw. On the backside it protrudes in a way that the electrical connection can be achieved. On the side a sample holder which can accommodate a normal SEM stub is fixed with two screws and can be varied in height. The other tower consists of a xyz-Attocube stack and is placed adjacent to the sensor tower at a distance that allows the tensile specimen to be gripped on both sides. The substrate on which the specimen is fixed on the actuation side is mounted with conductive silver paste to a part which is fixed on top of the Attocube stack. All parts are machined from Aluminum 6000 (AlSiMg). The whole setup can be mounted with only two screws, whereas for the Hysitron setup six screws are needed.

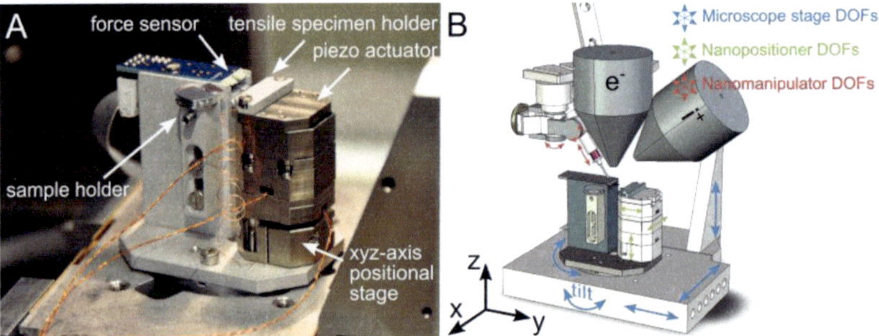


Figure 2.7: Femto Tools nanomechanical testing setup inside the vacuum chamber of the dual-beam microscope. (A) shows an optical image of the assembly with the labeling of all crucial parts. In a 3D-CAD drawing (B) the degrees of freedom for all moveable parts are highlighted.

The manipulator can be inserted on the bridgemount as described in the Hysitron setup in chapter 2.1.2, however this will limit the rotational degree of freedom since the sensor tower can collide with the manipulator if rotated too far. A roof mount was designed which allows fixing the Kleindiek manipulator on the roof of the microscope vacuum chamber in several positions. The testing setup can now be rotated 360° without collision and the limiting factor now is only the length of the cables. Since the manipulator is no longer connected to the tilting part of the stage, also in-plane misaligned specimens can be harvested by tilting the entire setup until the desired specimen is perfectly aligned. Another advantage of this setup is that it facilitates electron backscatter diffraction (EBSD) measurements during testing. For insertion of the EBSD detector the setup has to be rotated by 90° and also may necessitate tilting to 70°, which is never possible with the Hysitron design specified above. Most parts of the setup are therefore equipped with slotted holes to be displaced if shadowing effects need to be circumvented. The rotational degree of freedom (Figure 2.7B) between the two grips of the specimen is no longer available, but is also not needed for tensile testing, since precise alignment can be obtained by fine motion of the manipulator and the positioners. With this setup displacement is applied by scanning of the upper ANPx101-piezo. Maximum deflection and hence maximum displacement of 5 µm is given by the maximum voltage (100 V) that can be applied. The resolution results from the smallest increment in voltage, which the controller is able to apply. A LabView program which controls the deflection of the piezo was written and integrated into the program for data acquisition provided by the FemtoTools company. In this customized code, which can be found in Appendix A.1.4, parameters can be changed online, generating a powerful tool for unique investigations since displacement can be stopped, continued, reversed or even changed in rate on-the-fly.

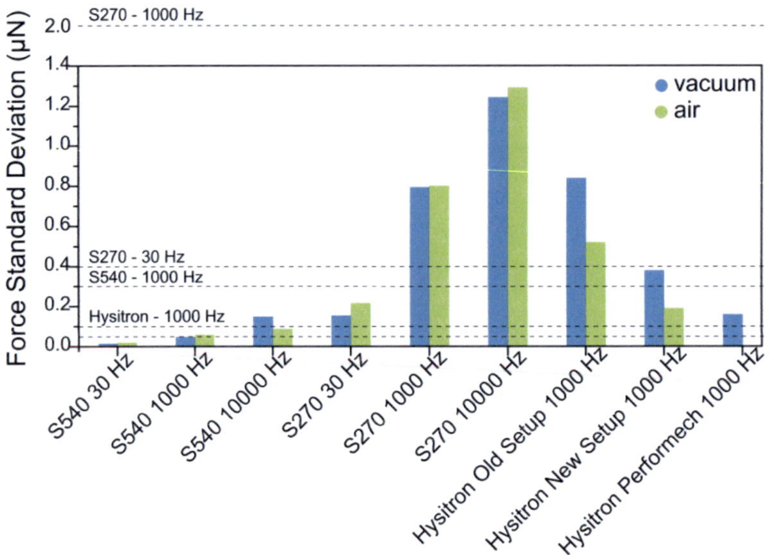



Figure 2.8: Force standard deviations of the different systems in air (green) and in vacuum (blue). The specifications given by the manufacturer are marked as horizontal dashed lines. For all DAQ rates the FemtoTools sensors are exceeding their specifications, whereas the Hysitron transducer never meets the specified values. The specification for FT-S270 at 1000 Hz is shown above the graph, since it is exceeding the range.

In order to compare the two systems, the noise floor has been analyzed for this setup using the same parameters as with the Hysitron setup. Recording time was always 10 seconds and DAQ rates of 30, 1000 and 10000 Hz were used to also compare the results to the specifications given by the data sheets of the manufacturer. As before, 5 tests were performed for each configuration and the presented results are average values of those. Figure 2.8 shows the comparison of the force noise floors for the two Femto Tools sensors and the Hysitron transducer. It can be seen that even if a DAQ rate one order of magnitude higher than reported in the specifications is applied, the Femto Tools force sensors are still significantly below the noise floor specified by the manufacturer. Also the noise level (FT-S540) is four times lower than the Hysitron device if the same DAQ rate is used. The difference between operation in vacuum and air is not as pronounced. The relatively low noise of this system is due to the rigid design configuration and the stiffness of the sensors themselves, having resonance frequencies of 2000 Hz (FT-S540) and 6400 Hz (FT-S270).

## 2.2 *In Situ* Tensile Testing Procedure

If the dimension of specimens is decreased to the nanoscale, tensile testing experiments inside the vacuum environment of an electron microscope are harboring several difficulties, which will be discussed in this chapter. Besides the manipulation, transfer and alignment of structures having diameters in the nanometer range, the gripping of specimens has to be accomplished. Furthermore, accurate and readily interpretable testing requires techniques for cross-sectional measurement and the structural characterization of the specimens. In order to obtain all mechanical properties digital image correlation (DIC) is used to determine the strain in the specimen.

### 2.2.1 *Tensile Test Preparation*

For the execution of a successful tensile test, several efforts have to be undertaken. Nanoscale specimens are often dispersed or grown in a high density, so that harvesting of a single nanowire is non-trivial. The scope is placed on extracting a single nanowire and the subsequent tensile testing with the Hysitron transducer, also shown in Figure 2.9. Once inserted into the vacuum chamber, the forest of nanowires (Figure 2.9A) is scanned for a suitable specimen. Since the microscope can only provide 2D information the setup has to be tilted in positive and negative direction to check the alignment in the tensile (focal) plane. The root of the nanowire is then approached by a combination of stage and manipulator motion. To exclude damage the final contact is achieved using the fine motion (cf. chapter 2.1.1) until: (i) a change in contrast occurs due to electrical contact or (ii) the specimen moves slightly or (iii) vibration of the specimen changes due to the newly obtained boundary conditions. The nanowire is then "glued" to the tip of the manipulator employing local e-beam induced deposition (EBID) of a solid Platinum (Pt) based metal-organic precursor, which is vaporized and delivered through an injection system and is then locally decomposed by the electrons.

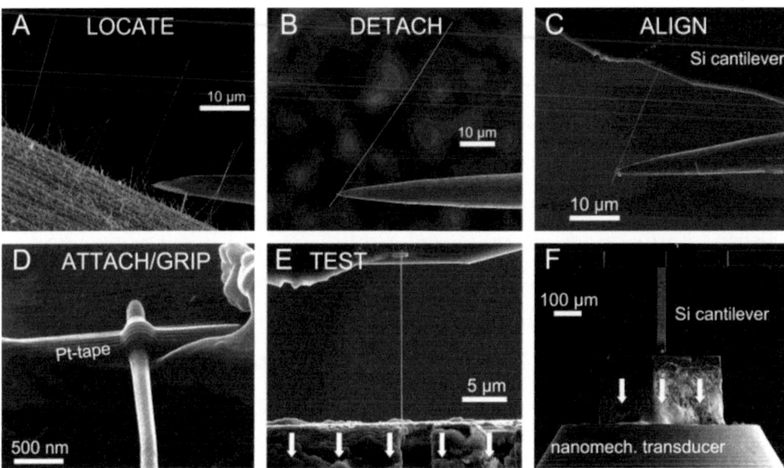


Figure 2.9: SEM micrographs showing harvesting of single nanowires. First the nanowire is located and approached (A), then it is gripped and detached from the substrate by the manipulator (B). The alignment to the cantilever is shown in (C). A zoomed image of the Pt "tape" used for gripping can be seen in (D). When the wire is attached on both sides the tensile test is performed by the actuation of the transducer (E). (F) shows an overview of the two bearings consisting of transducer tip and Si cantilever. The nanowire is clamped to the gap in between.

Since the nanowire is gripped close to its root small movements of the manipulator will cause stress concentrations at the bottom and detach the wire from the substrate. The tested portion of the wire being the free end is therefore stress-free and without any predeformation. Figure 2.9B shows the nanowire attached to the manipulator after detachment. As an alternative the manipulator tip can be attached to the distal end of the nanowire, which is then cut free at the substrate by the ion beam. However this method is avoided when the i-beam is expected to influence the mechanical response of the specimen. The nanowire is then aligned to the first gripping surface (Figure 2.9C), using the motion of the microscope stage and the manipulator. Once the contact is established, EBID is again employed to assure a persistent connection. This step has to be addressed carefully since the alignment to the first crosshead strongly determines the alignment of the nanowire to the force measurement and actuation axis of the setup and hence ensures a clean and uniaxial test. Given the difficulty of correcting for misorientations out of the image plane, AFM cantilevers can be used as a bearing, providing ample in-plane stiffness while also exhibiting some out-of-plane compliance. The success of this approach is on the one hand strongly depending on the degree of misalignment out-of-plane and on the other hand on the stiffness of the material being tested. The specimen is then cut free from the manipulator with the FIB using small beam currents ($\leq$10 pA) to avoid damage. The cut area of the specimen is never included into the gauge section of the specimen. Now the manipulator is retracted and the nanowire is aligned with its tensile axis parallel to the tensile direction of the setup by the rotation of the microscope stage. Afterwards the sample is driven to the transducer tip by the xy-Attocube stack and the z motion of the microscope. When in contact to the second crosshead, local EBID is used once more to grip the specimen (Figure 2.9). The nanowire is then ready for tensile testing, as can be seen from Figure 2.9E and F.

Displacement is applied by applying a DC voltage between the plates of the transducer for the first setup (cf. chapter 2.1.2) or by scanning the piezo of the Nanopositioner for the second setup (cf. chapter 0). Force and displacement data are recorded at rates of 100 to 500 Hz. The engineering tensile stress is obtained by dividing the force by the initial cross section of the specimen. Various routes to investigate the cross-sectional areas are mentioned in the particular chapters. However, the inherent compliance of force sensors and gripping modes used at the nanoscale exacerbates their usage for clean displacement measurement and the subsequent transformation into strain, provoking inaccuracy. Therefore digital image correlation is used as a non-contact method which provides accurate and direct measurements of local displacement/strain, (i) not necessitating contact with the specimen and (ii) providing sufficient spatial resolution to measure locally (cf. chapter 2.2.2).

After testing the location of fracture is used to determine whether a test is valid or not. Fracture at one of the grips indicates misalignment. Data from such a test requires further analysis, since due to the stochastical distribution of defects in small volumes a fracture close to the grips can also be caused by a weakest link instead of misalignment.

It is evident from Figure 2.9D, that the EBID employed for gripping of both ends of the wire provides a conformal coating and attachment of the specimen to the substrate. Although there is limited information available about the mechanical properties of this compound, it is known that the microstructure and the corresponding properties strongly depend on the deposition parameters. This method has proven to be sufficiently strong for specimens of several µm in diameter and forces up to 5 mN [95]. All depositions are performed using the electron beam as the source of dissociation. The energy is kept at a constant acceleration voltage, adjusting the beam dwell-time, beam current and pattern area according to the material and dimensions of the specimen.

If the specimen exceeds the dimensions of ~300 nm in diameter and has high strength, the regular Pt bond to a flat substrate is sometimes not strong enough resulting in slip at one of the grips. This is mostly due to an insufficient connection between the Pt tape on the specimen and the substrate. In these cases two methods of reinforcement have proven to enhance the strength of the connection: (i) FIB milling of a trough in which the wire is placed (see Appendix A.3.1). The Pt-organic deposit then seeps around the whole diameter providing the largest area and therefore the strongest connection possible, or (ii) deposition of multiple strips perpendicular to each other at angles of 45° to the tensile direction to obtain maximum resistance of the grips in shear [86].

### 2.2.2 *Image-Based Strain Measurement*

The quality of the results on mechanical testing not only hinges on the capability of resolving and measuring small forces, but also on the measurement of local displacement and the deduction of strain. Utilization of conventional displacement sensors, designed for macroscale testing (e.g. extensometers or strain gauges), is hard to realize and in most cases strictly impossible owing to the fact that nanoscale testing requires very high resolution and only allows for much smaller space in the setup. Also the inherent compliance of force sensors and gripping modes used at the nanoscale exacerbates their usage for clean displacement measurement and the subsequent transformation into strain, provoking inaccuracy. Hence a non-contact method which provides accurate and direct measurements of local displacement/strain is desired, (i) not necessitating contact with the specimen and (ii) providing sufficient spatial resolution to measure locally.

Digital image correlation has become a standard tool in the class of non-contacting methods. Random patterns can be employed to compare sub-regions throughout the images resulting in full-field deformation and/or motion measurement with sub-pixel resolution. Although white-light microscopy has been the predominant approach, DIC technique has been recently spread to electron microscopy.

In this work a custom software suite, employing the mathematical computing environment MATLAB® as the engine for calculations [153], developed by Eberl, Gianola and Thompson at Johns Hopkins University in Baltimore is used. The program which is an open-source package can be downloaded at a public domain [154]. The following paragraphs will provide a brief description of the method and the tools required to facilitate accurate strain measurements for the tensile testing of nanowires. A precious survey on DIC accompanied by a deep insight into the mathematical operations and the challenges beyond was recently published by Sutton [155].

DIC is usually performed by calculating the maximum correlation between consecutive images or subsets of images which are shifted with respect to each other [153]. To do so a relative stable contrast between consecutive images and a good signal-to-noise ratio are required. In this software an iterative approach is employed to maximize the 2D correlation coefficient by using a correlation algorithm provided by MATLAB®. The correlation coefficient $r_{ij}$ is given by:

$$r_{ij}\left(u,v,\frac{\partial u}{\partial x},\frac{\partial u}{\partial y},\frac{\partial v}{\partial x},\frac{\partial v}{\partial y}\right) = 1 - \frac{\sum_i \sum_j [F(x_i, y_j) - \bar{F}][G(x_i^\star, y_j^\star) - \bar{G}]}{\sqrt{\sum_i \sum_j [F(x_i, y_j) - \bar{F}]^2 \sum_i \sum_j [G(x_i^\star, y_j^\star) - \bar{G}]^2}} \qquad (16)$$

*F(x,y)* is the gray value or intensity at the point of interest *(x,y)* in the undeformed image, whereas *G(x*,y*)* is the gray value at a point *(x*,y*)* in the deformed image. *F* and *G* represent the mean values of the intensity matrices *F* and *G*. The coordinates *(x,y)* and *(x*,y*)* are related by the deformation which occurred between the two images [156]. If the motion of the object is perpendicular to the optical axis of the camera, the relation is given by [156]

$$x^\star = x + u + \frac{\partial u}{\partial x}\Delta x + \frac{\partial u}{\partial y}\Delta y \qquad (17)$$

$$y^\star = y + v + \frac{\partial v}{\partial x}\Delta x + \frac{\partial v}{\partial y}\Delta y \qquad (18)$$

where $u$ and $v$ are the translations for the subset centers in the $x$ and $y$ directions. The distances from the subset center to point $(x,y)$ are given by $\Delta x$ and $\Delta y$.

The quality of strain measurements with DIC strongly depends on the contrast which is needed for correlation. This simply means that not all grey value derivatives must be zero, because motion estimation is not possible in regions of constant grey values. Intensity variations in digital images can be of natural origin or also be introduced intentionally by surface decoration or machining. Several approaches for the introduction of surface contrast in SEM images and correction schemes to address imaging artifacts and aberrations were published by Sutton *et al.* [155]. Since these methods are generally used for larger dimensions, none of the proposed methods could be applied by concurrently avoiding damage of the nanowire or load bearing by the decoration itself. Therefore the natural contrast, which is strongly dependent on the growth method of the specimen, has to be used for image processing. Figure 2.10 shows intensity maps for three different subsets of images. Due to the pure crystallinity of the wires they usually do not provide enough contrast for image correlation. It can be seen in Figure 2.10A that there is little intensity variation in the gauge section of the nanowire along the tensile direction. However in the region of the grips a better intensity variation is obtained, mainly owing to the differences of the materials (bearing, wire, deposit) and the undulated topography in this area (Figure 2.10B). In some cases nanowires are also naturally decorated. This is mostly seen for specimens grown by VLS, where the wires are grown in higher densities and left over catalyst particles adhere to or new smaller wires emerge from the surface. As can be seen in Figure 2.10C the intensity variation along the tensile direction is much more pronounced for these specimens. Also EBID can be used to generate contrast along a nanowire but can also complicate interpretation due to the contamination discussed in the chapter above.

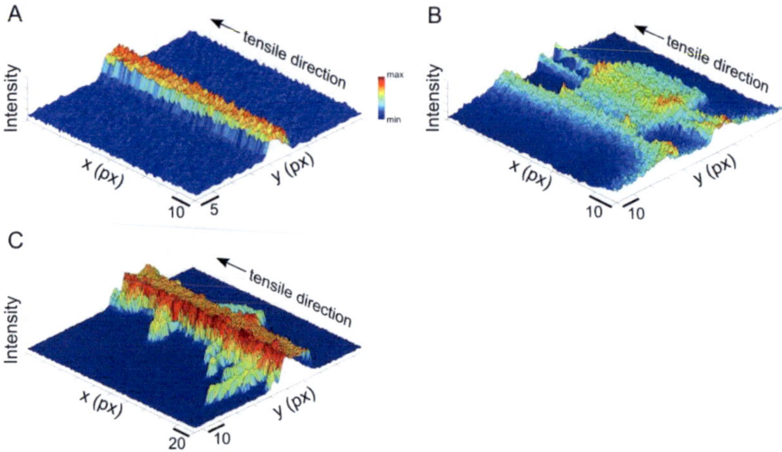

Figure 2.10: Intensity plots for different subsets of SEM images. (A) shows the intensity along the gauge section of a clean nanowire. In the tensile direction little variation in intensity is visible. In (B) the intensity plot of the Pt-organic deposited grip of the same wire can be seen, where more intensity variation is achieved. In (C) a much higher variation in intensity is visible in a gauge section of a naturally decorated nanowire, which facilitates thorough tracking along the length of the specimen.

The strain in the specimen is deduced by using the "1D average strain" tool in the program. In this method the horizontal displacement, which is the current position of the markers subtracted by the initial position of the same, is plotted versus the current position profile, which is the actual length of the specimen. Therefore the determination of strain in this program is always with respect to the current length, providing an Eularian frame (true strain) of reference instead of a Lagrangian frame (engineering strain), where mapping is performed with respect to the initial length of the specimen. Assuming proper alignment the axial strain from the full tensor is obtained. Strain is then gained by applying a linear least-square fit and measuring its slope between the two sets of markers or along all markers on the specimen respectively. If 2-point measurements using a set of markers on either side of the grip are performed, standard definitions like "engineering" or "true" strain become applicable. For linear elastic loading engineering and true strain deviate less than 5% with respect to each other at 10% total strain. Once the onset of plasticity occurs and the deformation localizes, the linear fit of all data will underestimate the local strain, which can be up to several hundred percent in these specimens. If the sample provides enough contrast, the localized region can be analyzed exclusively to approximate the local strain. The strain in this work is always referred to as "engineering strain" in all graphs, but should only be considered for homogenous deformation in the linear-elastic regime, whereas in cases of localization the term "mean total strain" is more appropriate.

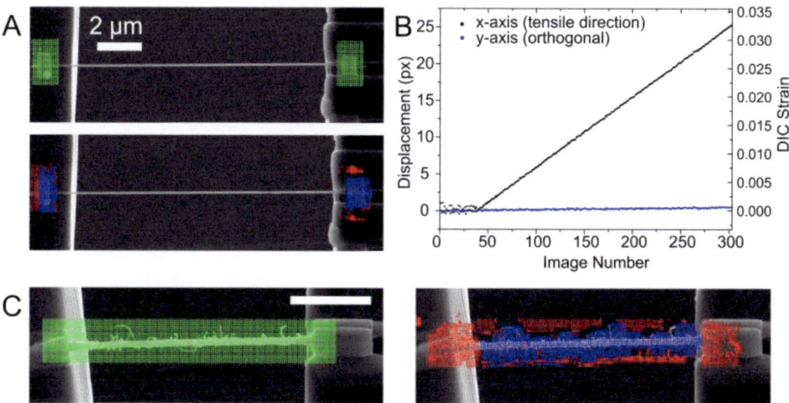


Figure 2.11: SEM micrographs with superimposed markers defined for image correlation (A). In the top image initial markers (green) are visible; the middle image represents all markers (red) short before fracture and the post-processed markers (blue). In (B) the horizontal (tensile direction) and vertical displacement and strain are plotted against the images. (C) shows the correlation of a VLS-grown wire, where strain mapping along the gage section is possible.

### *2.2.3 Contamination*

During imaging surfaces with the electron beam, carbonaceous and aqueous contaminations present in the vacuum chamber are deposited on the wire under the illumination with electrons. Also the metal-organic deposits, which are commonly used to grip nanowires, have been experimentally shown to delocalize and to decorate surfaces well away from the region of electron rastering. Gopal *et al.* [157] showed that the range of decoration is exceeding 10 µm on oxidized Si, which was assigned to the thermally assisted diffusion of Pt along the Si surface. We have largely observed the formation of contamination layers on nanowires. Additionally, a layer often already existed before the wires were heavily imaged. During imaging the thickness of the layer increased with time and was also found to be strongly correlating with the applied e-beam settings.

From Figure 7.2 and Figure 7.7 a smooth layer on the surface of the wires is visible, suggesting a core-sheath like structure. This can also be seen from Figure 7.8C where the wire is cut at the upstanding end with a low beam current. The contrast reveals the Au, which is heavy and hence brighter in the SEM micrograph. It is also visible that there is a thick layer of lower contrast, and therefore lighter, covering the Au wire. More detailed examinations of some of the fractured portions of wires in a ZEISS ULTRA*plus* electron microscope revealed that there is a surface layer that covers the whole wire (Figure 2.12A) and is partly removed (Figure 2.12B) by the insertion of oxygen that forms ozone which is subsequently used for etching. TEM investigations in Figure 2.12C show a wire, which is fully covered by a layer. This particular wire was heavily exposed to the electron beam performing tensile testing, EBSD and TEM. It can again be seen that the surrounding material is much lighter than the wire.

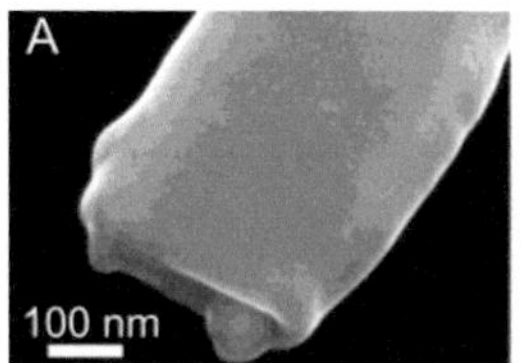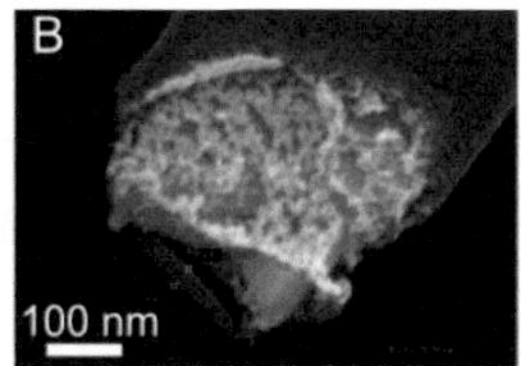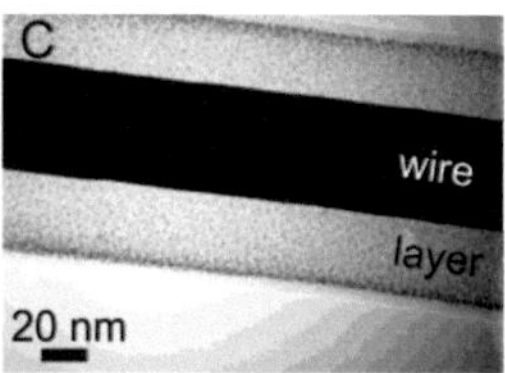


Figure 2.12: Surface morphology of tested Au whiskers; SEM images of the fracture side of a ribbon covered with contamination layer (A), and specimen after removing the layer (B). (C) shows a bright field TEM image of a wire after heavy imaging with the electron beam; the surrounding contamination layer is almost as thick as the wire.

A contamination layer on the specimen will affect the measurement of the true specimen dimensions and will possibly bear a considerably fraction of the applied load. Therefore it is inevitable to address the effect of the contamination layer on the mechanical response of nanostructures, which was accomplished by Gianola *et al.* [86]. With the experimentally obtained values for ion-beam induced Pt-organic deposition by Utke *et al.* [158] one can calculate the effect of a Pt-organic layer on the measurements of the Young's modulus. The following calculation is made under the assumption of uniform strain (Voigt composite) of an Au nanowire surrounded by a Pt-organic sheath. The effective modulus $E_{eff}$ can be expressed by the Young's modulus $E_{Pt-C}$ of the PT-organic, the Young's modulus $E_{NW}$ of the nanowire and the volume fraction $v_{PT-C}$ of the Pt-organic sheath.

$$E_{eff} = E_{Pt-C} v_{Pt-C} + E_{NW}(1 - v_{Pt-C}) \tag{19}$$

Assuming a cylindrical core shell structure, $v_{PT-C}$ can be expressed as

$$v_{Pt-C} = \left(\frac{t}{d}\left(1 + \frac{t}{d}\right)\right) \Big/ \left(\frac{1}{4} + \frac{t}{d}\left(1 + \frac{t}{d}\right)\right) \tag{20}$$

where $t$ is the thickness of the shell and d is the diameter of the nanowire. Assuming a modulus ratio $E_{Pt\text{-}C}/E_{NW} = 0.1$ [158], the effective modulus of the composite structure would change by more than 10% if $t/d > 1/33$.

Following this rationale, the accuracy is demonstrated on the gold nanowire from Figure 2.12C. For the contamination the value after tensile testing is used ($d_{nom} = 78$ nm). It has to be noted that due to heavy electron imaging during EBSD and TEM the contamination in the TEM image is much stronger than during the tensile experiment. The diameter of the wire is measured to be 48 nm, leading to a layer thickness $t = 15$ nm. If the effective modulus, which is determined to be 38 GPa (Figure 2.13A), is inserted into equation 20 the Young's modulus of the Au nanowire is calculated to be 86 GPa. With less than 5% deviation from the calculated <110> directional modulus this value is in the range of experimental uncertainty and confirms the validity of this simple approach. Figure 2.13B emphasizes the effect of a contamination layer on the Young's modulus of these small samples. For three different diameters the modulus is plotted versus the layer thickness showing increasing impact with decreasing wire diameter.

With this method the thickness of possible layers on the specimens can be approximated and the mechanical data can be corrected if the elastic properties of both, the core (nanowire) and the shell (layer) are known.

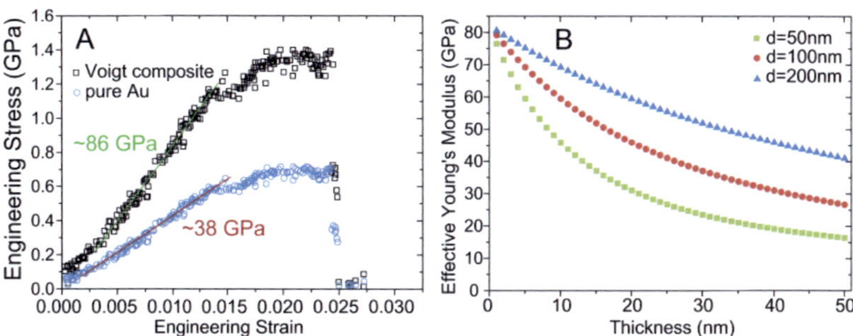

Figure 2.13: Influence of Pt-organic and other contaminations on the measurements of Young's modulus. Measurements assuming pure Au yield a modulus of 38 GPa (A). If a core-sheath structure is accounted for using equation 19, the modulus is 86 GPa which is well in the range of experimental uncertainty. (B) emphasizes the increasing influence on experimental error with increasing layer thickness and/or decreasing wire diameter.

# 3 Molybdenum Fibers

As a proof of concept Mo fibers were investigated under tension. These fibers are grown by directional solidification of a Ni-Al-Mo eutectic. By subsequently etching the matrix the fibers are obtained. Mo is a body-centered cubic transition metal. Under ambient conditions Mo forms a passivation layer. The bulk elastic modulus of Mo is around 320 GPa and yield strength is in the three digit MPa regime. This work was performed and published [159] in close collaboration with Kurt Johanns at Oak Ridge National Laboratory, Knoxville, Tennessee. The same group also investigated micropillars etched from the same material, so that a direct comparison between the results is possible.

## 3.1 Tensile Testing of Molybdenum Fibers

The Mo fibers were tested in tension according to the procedure described in chapter 2.2. Since these fibers are larger than nanowire specimens, the trough cutting technique had to be employed for every single specimen to achieve a strong fixation. This technique is further described in A.3. A total of 19 as-grown, 8 prestrained by 4% and 11 prestrained by 16% fibers were investigated at slow quasi-constant rates of $10^{-4}$/s. The fibers had width of 300-600 nm and lengths of 9-41 µm. For simplicity the batches of fibers with different amounts of prestrains are noted as E plus the amount of prestrain (E0, E4, E16).

The recorded mechanical response showed distinct variations, which can be observed in Figure 3.1A. Generally, it can be divided into three different modes. Mode 1 (blue curve) is characterized by linear-elastic loading with abrupt failure at high stresses. Without any anterior sign of plastic deformation the fiber fractured within one frame of the acquired images. Correlations of the modes of behavior with the amount of prestrain are visible in the inset of Figure 3.1A. Mode 1 behavior was observed in 69% of the as-grown fibers (E0) and 31% of the 4% prestrained specimens (E4). It was never obtained for the 16% prestrained fibers (E16).

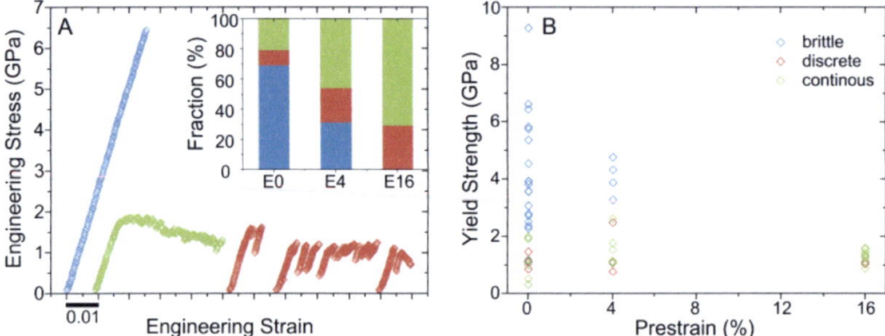


Figure 3.1: Three distinct modes of mechanical behavior are shown in (A), ranging from linear-elastic loading until abrupt fracture (mode 1, blue curve) to continuous plastic deformation without pronounced hardening (mode 2, green curve) or intermittent plastic deformation (mode 3, red curve) with stress drops and subsequent linear-elastic loading. The inset of (A) shows the occurrence of the respective mode of behavior per prestrain. In (B) the yield strength is plotted versus the amount of prestrain. The color coding correlates the yield stress to the mode of mechanical response.

In the second mode of behavior (green curve) the linear-elastic regime was followed by stable continuous plasticity with little or no hardening. Often the curve showed softening accompanied with the slow growth of a neck. Mode 2 behavior was found in 21% of E0, in 46% of E4 and 71% of E16. In a third mode of mechanical behavior (red curve) the linear elastic regime was followed by intermittent plasticity, i.e. plastic deformation was often interrupted by stress-drops and subsequent loading along the linear-elastic slope until plastic deformation recommenced. Mode 3 behavior was observed in 10% of the as-grown fibers, 23% of E4 and 29% of E16 showed this intermittent character. As can be seen in Figure 3.1B, mode 2 and 3 behavior occurred at significantly lower stress levels than mode 1.

Also the deformation and fracture behavior showed distinct variations. For all specimens the deformation was localized, i.e. it occurred at discrete spots and was not uniformly distributed along the fiber. In some cases the deformation was spreading along the fiber length and sometimes the nucleation of deformation sites at multiple locations was observed. Usually the thinnest region was the first to deform. Although mode 1 solely showed linear elastic loading until abrupt failure the fracture surfaces were not indicative of brittle fracture. Instead many of the fracture surfaces showed ductile features. Representative images for this mode of behavior are shown in Figure 3.2A and B, where Figure 3.2A is reminiscent of cleavage and Figure 3.2B shows shear fracture with small amounts of ductility. This morphology was never observed in E16 specimens. Figure 3.2C represents the observations of deformation and fracture for mode 2 behavior. After the onset of plastic deformation the fiber necked at this single spot until the two parts completely separated. Although this behavior was obtained for all three levels of prestrain the amount of necking varied from sample to sample and was not depending on prestrain. An exemplary SEM micrograph for the third mode of mechanical response is shown in Figure 3.2D. In this case the deformation usually started in one region and then propagated away from this site repeatedly activating new deformation sites, comparable to the formation of Lüders bands. In some occasions the propagation stopped and deformation took place at other locations. The stress-drops in the curve corresponded to the discontinuous propagation of the plastic zone. A larger amount of strain was achieved at this event, which was visible through the larger shear offsets in the SEM images (Figure 3.2D). The failure modes for this type of behavior were mixed between necking and shear failure with less plastic deformation. As in the other modes of mechanical response this type of deformation and fracture was observed independent of the amount of prestrain. Although the different modes of deformation occurred through all levels of prestrain, the frequency of each deformation and fracture behavior showed a strong correlation to the level of prestrain of the fiber.

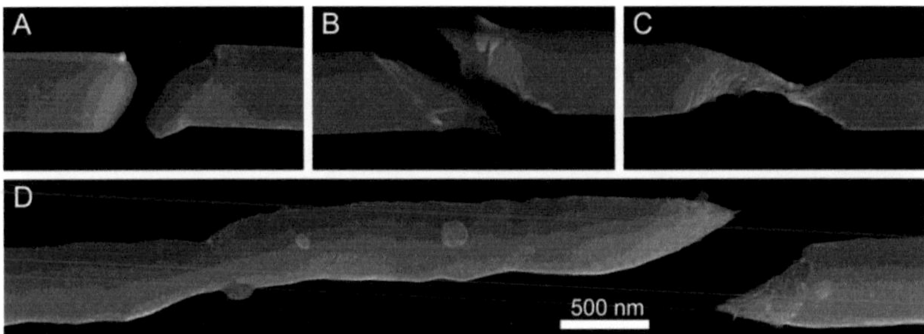

Figure 3.2: Deformation and fracture morphology of Mo fibers. (A) and (B) show the fracture behavior of mode 1. Although the mechanical response seemed to be brittle, the fracture site only rarely showed indications of cleavage (A) and many times shear fracture with small amounts of ductility took place (B). For mode 2 plastic deformation occurred at a single location and the fiber necked until final failure (C). Mode 3 exhibited discontinuous propagation of the plastic zone (D). The deformation propagated away and repeatedly activated new deformation sites. The discontinuous drops in stress were correlated to the formation of larger shear offsets on the fiber surface. Failure modes of this type of behavior were mixed between necking and shear.

In a few occasions the formation and propagation of classic Lüders bands could be observed. Figure 3.3 shows an example of a fiber with 4% prestrain. This particular fiber was strained consecutive times and the individual tests are highlighted in different colors. After the first test the fiber was unloaded, since the limit of the programmed displacement ramp was reached. In the second test the fiber fractured and was then regripped for the third test. The fibers exhibited an upper and a lower yield point at a relatively constant stress level, as can be seen in Figure 3.3A. Observation of the Lüders band propagation was sometimes alleviated by the existence of a surface layer that fractured and changed the contrast on the surface (Figure 3.3B). In this test the fiber exhibited a distinct upper yield point at 2.5 GPa. The stress then dropped to zero and subsequently the fiber was reloaded. Flow took place at a lower level of about 1 GPa with a high number of stress-drops during which the Lüders band extended along the fiber. In the second and third test the fiber immediately started to flow at the lower stress level after linear-elastic loading. Figure 3.3B shows three different SEM images taken during straining. It can be seen that the deformation front is propagating to one side.

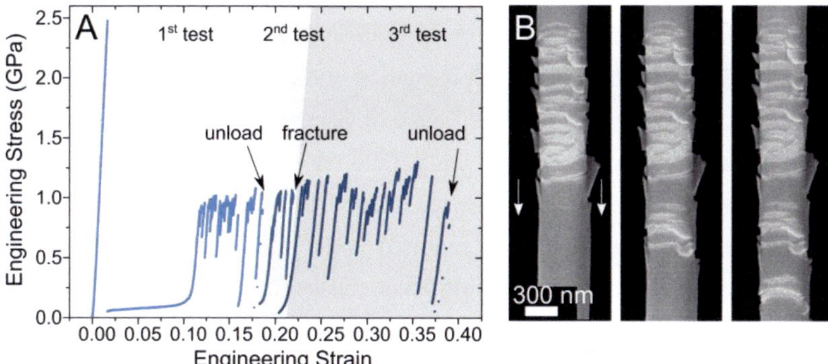



Figure 3.3: Extended plastic deformation in Mo fibers. In several occasions the specimens yielded at a high stress level and then flow took place at a lower level (A). The flow in (A) could be correlated to the propagation of a classic Lüders band along the fiber. Sometimes a layer existed that peeled off and hence changed the contrast, alleviating the *in situ* observations.

Figure 3.1B shows the yield strength as a function of prestrain. As can be seen the as-grown fibers are the strongest, but also exhibit a large scatter with strength varying between ~1 and 10 GPa. The weakest were the specimens with 16% prestrain slightly higher than 1 GPa and only little scatter. Fibers with 4% prestrain showed intermediate behavior with strength and scatter lying between the two aforementioned extremes.

## 3.2 Discussion

The fibers of different amounts of prestrain showed distinct mechanical responses and deformation morphologies, ranging from the two extreme cases of linear-elastic loading until sudden failure to yielding and continuous plastic deformation accompanied by necking at a single location. All modes of prestrain showed the low level of yield stress of ~1 GPa. However, the as-grown fibers exhibited the whole spectrum between this value and the theoretical strength of ~10 GPa, the 4% prestrained fibers showed intermittent stress-strain curves and a reduced but still significant scatter.

The linear-elastic behavior, mostly observed for the E0 specimens, is not of strictly brittle nature. Deformation morphology suggests at least some amount of plasticity. This behavior is owing to the poor feedback control of the nanomechanical controller, which is further discussed in A.1.3. For these comparably large specimens the effect is more pronounced than for the smaller nanowires reported later. As a consequence a huge forward surge of the tip at the yield point causes fracture instead of flow at a lower stress-level for most of the E0 specimens. Only for fibers exhibiting a lower yield stress, usually from the E4 batch containing few defects, is the feedback loop of the controller fast enough to not destroy the fiber. For these specimens flow takes place at a lower stress-level, as shown in Figure 3.3. The necking right after yielding in the continuous curves is identical to the behavior of highly work-hardened metals.

Regarding the strength of the fibers, the observations are similar to the compression experiments performed by Bei *et al.* [121]. The as-grown specimens reach the theoretical strength of ~10 GPa and the highly prestrained specimens show yield strengths of ~1 GPa. For both sets of experiments the scatter is reduced with increasing prestrain. In contrast to the aforementioned similarities there is one significant difference, which can easily be seen from Figure 3.4. The as-grown fibers subjected to tensile stress show large scatter, whereas the compressed specimens consistently show high strength at the theoretical limit. This marked difference is due to the fact that all the tested pillars are essentially dislocation free and the lion's share of the fibers is not. These findings are in agreement with the investigations of Phani and colleagues [160], who performed TEM on the same batch of fibers. They were able to show that even in the as-grown fibers, small groups of dislocations could be found exhibiting an average separation distance of ~37 µm. With this knowledge the observed difference in scatter can be explained by a weakest link argument. Significantly larger gauge lengths of the tensile fibers (9-41 µm) bury a much higher probability of finding a critical defect, which then weakens the entire fiber, in comparison with the short pillars (~3 µm). This stochastic event is responsible for the observed scatter in the as-grown tensile specimens. In contrast, the short pillars are all free of defects and consequently yield at the theoretical limit. For comparison, the volume tested in a 30 µm long specimen tension is 10 times as much as the volume of one specimen tested in compression.

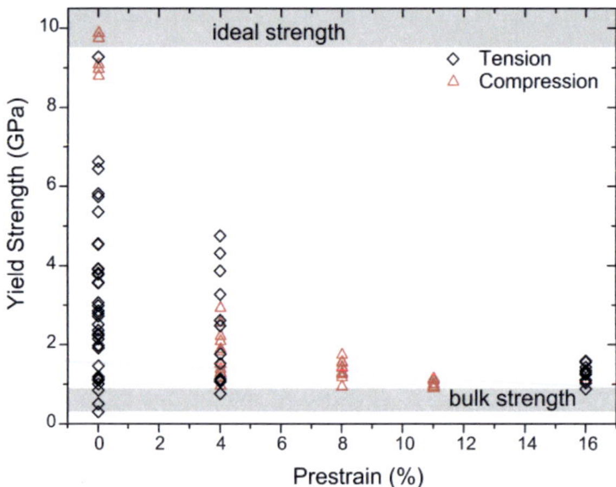


Figure 3.4: Comparison of results of tensile testing with compression [121] for various prestrains. In contrast to the compression the as-grown fibers tested in tension show tremendous scatter in yield strength. 11% and 16% prestrained specimens show deterministic bulk-like behavior. With decreasing prestrain the scatter is increasing.

In the context of weakest link statistics the dependence of strength can be shown by plotting the measured yield strengths versus the length, surface area or volume of the specimens to account for the difference in probability of finding a defect. In Figure 3.5A the tensile specimens do not show this dependence but a large scatter, calling for a larger data pool for confirmation. It should be noted that also the character of the existent defect is important, since a dislocation having a Schmid factor of zero will not influence the mechanical response. Furthermore, in Figure 3.5A the yield strength as a function of gauge length is compared to the results for pillar compression [121] and Monte Carlo Simulations for larger gauge lengths [159]. Short tensile specimens only show high strengths at the theoretical limit. With increasing length the scatter increases significantly and exists for a large range. The simulations of Johanns *et al.* [159] predict a transition to deterministic behavior at the strength of bulk material first around 1000 µm. The dependence on specimen length can also be proven by testing a single fiber consecutive times, as depicted in Figure 3.5B. An as-grown fiber that contained some dislocation was tensile tested and after fracture one of the portions was regripped and tested again. This time not only the stress-strain behavior was different, but also the strength was significantly higher, since the defect that weakened the specimen was eliminated in the test before. If the same is performed with E16, the mechanical response of the consecutive tests is nearly identical to the first test because of the high and more homogeneous initial dislocation density.

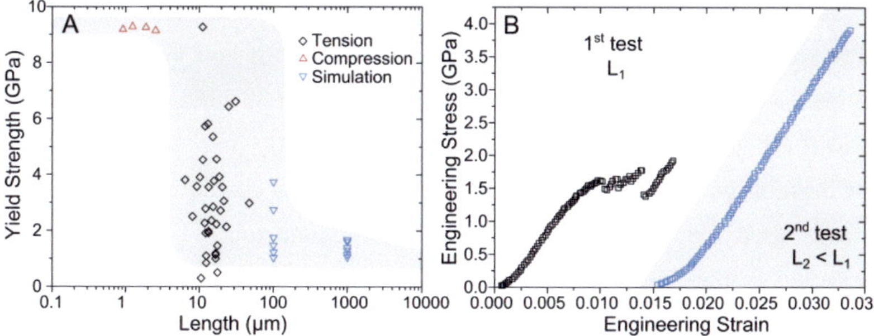

Figure 3.5: Probabilistic treatment of the results for as-grown Mo fibers. The yield strength in tension is compared with compression [121] and simulations [159] (A). The scatter is increasing between the two deterministic extremes for high yield stresses in short fibers and long fibers at the strength of bulk material. Consecutive tensile testing (B) of a single as-grown fiber confirms the probabilistic argument. A fiber that contained some defect shows higher strength and a distinct mechanical response once this weakest link is eliminated.

The dislocation density in the highly prestrained specimens is so large that also the smallest specimens contain a high number of dislocations. As a consequence the behavior and the strength approximate to that of the bulk material with only low scatter. The bottomline around 1 GPa (Figure 3.4), which is found in all tensile batches is taken as the bulk value here. However, it is possible that the bulk strength of E0 is less than in the prestrained samples, since the few dislocations are able to move relatively easy. In the fibers with high dislocation densities other dislocations can obstruct the path and therefore require higher stresses to facilitate plastic deformation. In the E4 specimens with medium dislocation density plastic deformation can move along significant parts of the gauge section after yielding. Once a dislocation is obstructed by another defect, plastic deformation is continued at the next weaker site or deformation is terminated by fracture. This is also valid for E0 but only observed rarely due to the insufficient control of the transducer at high stresses (cf. A.1.3).

The Young's modulus measurements of the tensile experiments reveal large scatter and a mean modulus that is significantly too low throughout all levels of prestrain (E0: 215±74 GPa, E4: 190±38 GPa, E16: 192±61 GPa) if compared to the calculated modulus in <100> direction (360 GPa). For these large specimens it is possible that there is co-deformation of the Pt tape used for fixation (cf. A.3.1). To elucidate this, Pt markers were deposited in the gauge section and subsequently the strain was measured independently of the grips. DIC returns 295 GPa between the two grips, including possible strain reduction by the Pt markers in the gauge section. The modulus measured between the two markers is measured to be 320 GPa. This shows that at high load and large sized specimens there is some compliance of the grips. Therefore the measured strains have to be regarded with care. The strong scatter may be caused by the unique character of deposition for each grip. Despite that other factors could influence the measurements, which comprise a contamination layer (chapter 2.2.3), a native oxide laxer, a residual layer of the etched matrix or misalignment, leading to corruption of strain measurement. The influence of an oxide layer and misalignment of the specimen will be revisited in chapter 5.3 of Si nanowire tensile testing. Nevertheless, the good agreement between the tensile tests presented in this work with the previous pillar tests on the same material validates the newly developed setup.

# 4 Vanadium-Dioxide Nanowires

Vanadium-Dioxide nanowires were grown and strained in tension in combination with electrical measurements. Pure strain-free $VO_2$ is subjected to a phase transformation at 341K from the high-temperature tetragonal metallic R phase to the monoclinic insulating M1 phase, accompanied by a change in resistance of several orders of magnitude, commonly known as metal to-insulator transition (MIT) [150]. Additionally, the presence of an intermediate M2 phase has been observed under uniaxial stress or from doping of $VO_2$ [161]. Pouget *et al.* demonstrated that the M2 phase could be stabilized by uniaxial stress along $[110]_R$ or $[1\text{-}10]_R$ and also postulated this metastable M2 phase to be a Mott insulator [161]. Hence $VO_2$ experiences an insulator-to-insulator transition (IIT) [162] if strained uniaxially.

## 4.1 Tensile testing of Vanadium-Dioxide Nanowires

The $VO_2$ nanowires studied in this work were grown in a home-built tube furnace. An alumina crucible contaminated with $VO_2$ powder was placed in the furnace and oxidized Si (100) wafers were positioned in the vicinity of the powder. The furnace was heated to 800°C and kept there for 2 hours at a constant flow of Ar at 800 sccm. This leads to the growth of individual nanowires by vapor-solid (VS) growth, which is assisted by the oxide of the Si wafer. Detailed information on the growth and the associated mechanisms can be found elsewhere [163].

The as-grown nanowires were characterized in the SEM in conjunction with energy-dispersive X-ray spectroscopy (EDX) and EBSD. Characterization by SEM showed square or rectangular cross sections with the highest edge ratio observed to be 2:1. Edge lengths ranged from 50 to 600 nm and the overall length of the nanowires was between 5 and 50 µm. Figure

4.1B shows a SEM micrograph with several wires reaching over the edges of the sample. In contrast to the observations of others, growth was limited to regions close to the edges of the substrate. The inset of Figure 4.1B shows a high resolution SEM micrograph of the faceted surface of an individual wire, foreshadowing the high crystalline quality of the specimens. EDX analysis confirmed the nanowires to be free of impurities at higher concentrations. In EBSD all wires were found to be single crystalline and to be in the monoclinic M1 phase, all with the same [001] axial growth direction.

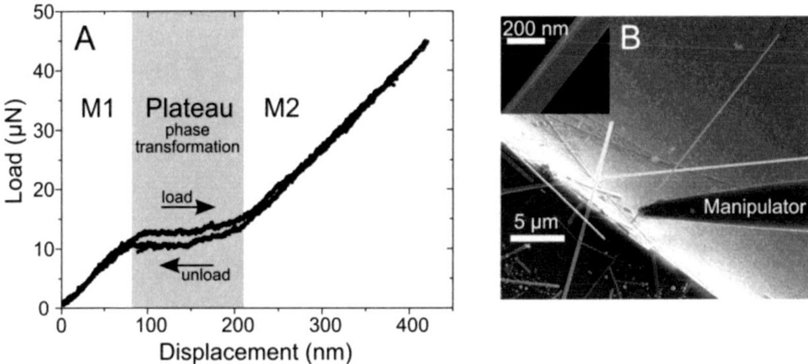


Figure 4.1: Load-unload curve (A) showing the superelasticity of VO2 nanowires. The curve shows the phase transformation from insulating M1 phase to insulating M2 phase upon mechanical straining. If the wire is unloaded the superelastic behavior of the material becomes evident. (B) shows VO2 nanowires growing along the Si substrate, which allows for easy harvesting.

Tensile testing was performed at quasi-constant strain rates of $10^{-4}$/s on 10 nanowires. During straining, all wires showed linear-elastic behavior until a change in slope could be detected and a plateau was visible, which can be attributed to a phase transformation from the insulating M1 phase to the insulating M2 phase. Upon further straining, the slope changed again and the wires were loaded linear-elastically until failure. Figure 4.1A shows a representative load-displacement curve, where different phases and the plateau are highlighted. It can be seen that the transformation is fully reversible, showing only a slight hysteresis at the plateau. During variation of the strain-rates between $10^{-4}$/s and $10^{-2}$/s, no differences in mechanical response were found. At stresses around 0.5 GPa and average strains of 0.6% the slope of the curve changed to zero and the plateau began. After ~0.5% of further straining (average plateau strain) the slope changed again and linear-elastic loading commenced until brittle failure. In most cases the wire broke off completely not leaving fracture surfaces to be investigated. In cases where a wire was retained, one of the fractured portions was regripped and pulled again until fracture, resulting in an increase in strength in all cases.

For the wire of Figure 4.2A, which was tested three times, all curves are shown. It can be seen that the tensile strength increased significantly. Figure 4.2B shows the characteristic fracture surfaces of a wire that are oriented perpendicular to the wire axis showing no hint of plastic deformation. Nominal fracture stresses of all pristine wires were found to lie between 2.5 and 7.1 GPa, the total strains were between 3.3 and 6.0%. If plateau and fracture stresses are plotted versus the edge lengths of the nanowires (Figure 4.2C) it can be seen that the plateau stress is consistent around 0.56±0.13 GPa and does not depend on size. The fracture strength increases only slightly with decreasing edge length, although there is significant scatter. The Young's moduli measured before and after the plateau can be seen in Figure 4.2D. In both cases the linear regimes without the non-linearities that appear at the onset of straining were used. With the exception of one specimen, the modulus of the M1 phase was found to be lower than that of the M2 phase, the average values are 100±32 and 120±25 GPa, respectively.

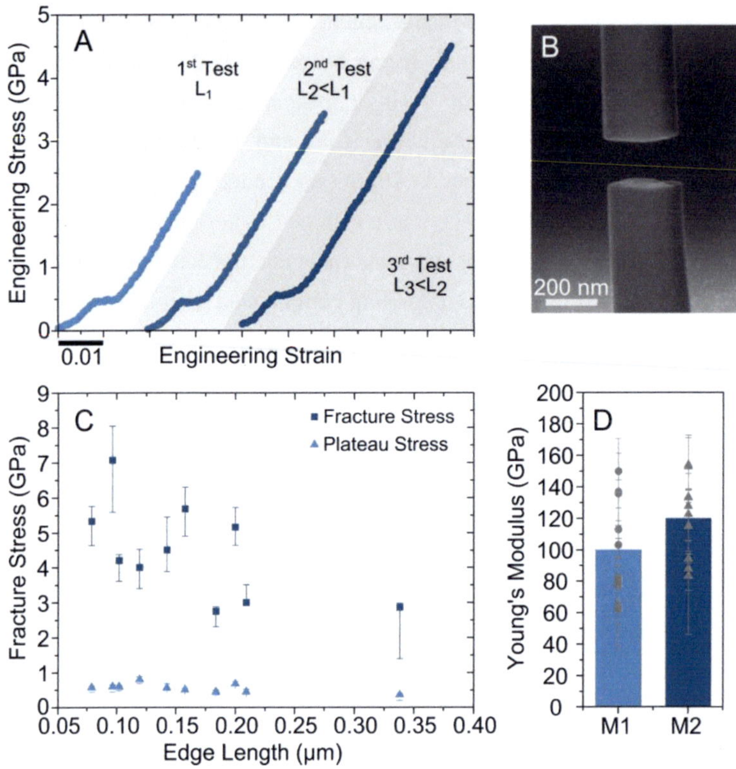



Figure 4.2: Stress-strain curves (A) of tensile tests show a plateau around 0.5 GPa during phase transformation. Three consecutive tests were performed and with each test the length of the wire decreased and the wire became stronger, fortifying the weakest-link behavior of this brittle material. In (B) the brittle fracture surface perpendicular to the tensile direction of a tested specimen can be seen. During all tests no signs of plasticity were observed. In (C) the fracture stress and plateau stress of all tested specimens is plotted over their minimum size. The Young's Modulus of both phases, before and after the plateau, is shown in (D). The columns represent the mean average values for each phase.

Additionally to the mechanical tests electrical measurements were performed. A schematic of the setup is shown in Figure 4.3A. The nanowire was gripped by EBID of Pt between the Kleindiek MM3A manipulator, which is also used for harvesting, and a piezo actuator equipped with the same W probe on the other side. Displacement was applied by the piezo actuator and the corresponding tensile strain was measured by digital image correlation (DIC). Using a Keithley 2400 series source meter, a constant voltage was applied between the two tips and currents as low as several nanoamperes were measured. In this two point measurements, a voltage between 0.2 and 2.0 V was applied in steps of 0.1 V until a current could be detected. The wire was then strained and the current vs. time as well as the scanning electron microscopy images were recorded.

Figure 4.3B and C show the resistance from the electrical measurements plotted versus time and tensile strain respectively. A constant strain rate was applied, and micrographs were taken continuously. In the image sequences, it was observed that the wires were usually bent slightly before tensile testing (i). During initial displacement the wires were straightened, also with respect to preexisting misalignment (ii). When the wire was straight, further straining led to an increase in resistance (iii). At a certain point, a strong increase in resistance was observed (iv) and concurrently the wire was slightly oscillating in the images. After about 0.4% of further straining and a total increase of about 255 k$\Omega$, the increase in resistance slowed down, and the oscillations in the image stopped (v). Then, the resistance was increasing linearly, again (vi), but with steeper slope compared to regime (ii). When the strain was reversed, the resistance decreased in the same fashion as the increase had occurred indicating that the behavior is fully reversible. It can be seen that the strain of about 0.5% related to the phase transformation is consistent with the mechanical measurements.

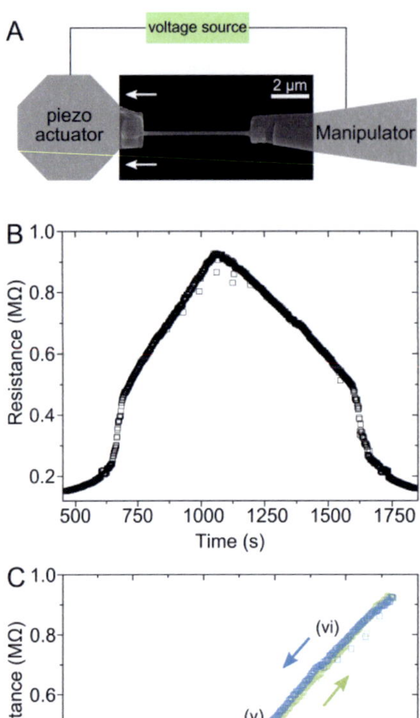


Figure 4.3: (A) Setup for the electrical measurements performed on single nanowires. The SEM micrograph shows the nanowire gripped between two probes. The left side is fixed to a piezo actuator which applies displacement. A constant voltage is sourced and currents as small as several nanoamperes are measured. Electrical measurements performed on a single nanowire, which was cycled two times with consistent curves. The resistance over time for one cycle is shown in (B). In (C) the resistance is correlated with the strain obtained by DIC. During the phase transformation a strong increase in resistance can be observed.

## 4.2 Discussion

The mechanical tensile curves of Figure 4.1A and Figure 4.2A suggest a first order phase transition upon uniaxial tensile straining from the monoclinic M1 to the insulating M2 phase, as has been discussed before [161]. In the mechanical data, the transition corresponds to a region in the stress strain curve where the Young's modulus drops to zero. In this region, the M2 phase nucleates and spreads out until the entire matrix is transformed and the modulus deviates again from zero. As can be seen from Figure 4.2C, there is little or no size dependence of the stress required for this transformation. Although there is large scatter in the strength, a trend for an increase in fracture strength with decreasing edge length is observed in Figure 4.2C. If the data is plotted versus length, volume or surface area, only weak correlations are found and from this a statement about the nature of defects cannot be made. The weakest link concept seems to be relevant since consecutive testing of a single wire shows rising strengths (Figure 4.2A) with decreasing wire length. This is due to the fact that fracture always takes place at the weakest link of the wire. Once this defect is eliminated, higher stresses can be reached, limited by the activation stress of the next defect.

Despite the new insights into a material that only rarely was tested under uniaxial tensile strain [162], there is one important point to the work of this thesis that can be shown with these data. This phase transformation material is dedicated for the verification of accuracy of the tensile setup. It exhibits a phase transition, which is independent of size and therefore the transformation stress acts as a proof stress for the stress measurements. As can be seen from Figure 4.2C, there is little scatter in the stress and little or no size dependence of the stress required for this transformation. However, the fracture strength, which was recorded in the same experiments shows significant scatter. Setups for nanomechanical testing are often very delicate and noise and scatter in experiments are common [104, 109]. The fact that there is considerable noise in the fracture strength and only little variation in the transformation stress strictly proves that the experiment itself is accurate and reproducible and that the scatter of the fracture strength is an intrinsic effect not caused by experimental uncertainties in the setup.

Resistivity measurements in a $VO_2$ nanowire have been performed for the first time. One concern with respect to the resistivity measurement is Joule heating due to the electrical current that is applied. Using the measured values of thermal conductivity of [164], we estimate Joule heating in a nanowire with edge lengths of 200 nm and an overall length of 5 µm to be less than 4 K for the most unfavorable assumptions. This indicates that thermal effects can be largely ignored in the electrical test.

The transformation from the M1 to the M2 phase is correlated by a strong increase in resistivity. In addition, in the electrical measurements, mechanical oscillations of the wire were observed at the phase transformation using SEM. This may be explained by the vanishing Young's modulus which leads to resonances in the setup, which was used for applying strain, here. These oscillations do not occur in the stiffer system for the mechanical tests.

For the M2 phase, the gauge factor, which relates the relative change in resistance to mechanical strain, was calculated to be 51 for this phase (1% of strain would lead to a relative increase in resistivity of 51%). Compared to other materials, this value is high but still lower than that of single crystalline silicon (-100...200, depending on doping) which is used in piezoresistive sensors. Also, the $VO_2$ nanowires could be used as mechanical switches in NEMS devices based on the large change in resistivity when a stress about 0.5 GPa is applied.

# 5 Silicon Nanowires

Si nanowires with <111> and <100> axial orientations were tested in tension. Si is the most common metalloid. It crystallizes in the diamond cubic crystal structure, like other group IV elements. Therefore each Si atom shares electrons with four other Si atoms, forming four single covalent bondings. The mechanical properties are anisotropic; depending on the crystallographic orientation the young's modulus varies between 130 GPa for <100> and 189 GPa for <111>. Si forms a native oxide layer of approximately 1-3 nm thickness. Depending on the growth method and the involved chemicals or catalysators the oxide layer for nanowires can be thicker.

## 5.1 Tensile Testing of <111> Oriented Nanowires

Tensile tests were performed on two classes of <111> oriented Si nanowires; the first was grown on an oxidized Si substrate and the second on silicon on insulator (SOI) (cf. chapter 1.6). Out of the first class 37 nanowires with diameters between 127 nm and 317 nm were tested. From the second 14 specimens with diameters between 65 nm and 239 nm were investigated. The tension was applied in feedback-enabled displacement control with nominal strain-rates of $10^{-4}$/s. Among 51 nanowires tested in total, all showed high strength in the range of 3 to 9 GPa. As can be seen from Figure 5.1, the mechanical response of both classes solely exhibited linear-elastic behavior with no signs of plastic deformation. The fracture strain of the tested specimens varied between 3 and 8%. From the highlighted stress-strain curves it is visible that no strong size dependence on strength could be observed.

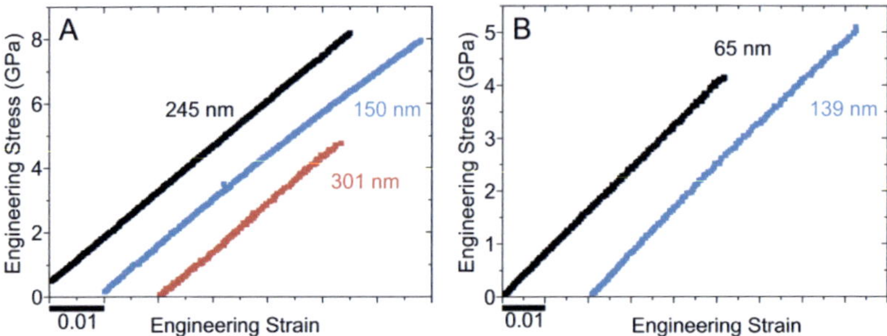

Figure 5.1: Stress-strain curves for <111> axial oriented Si nanowires grown on oxidized Si substrate (A) and SOI (B). Fracture at high stresses solely showed linear-elastic behavior with no signs of plastic deformation. No strong dependence of fracture stress and size could be observed.

The nanowires were strained at quasi-constant strain rates. Due to the surface irregularities of some of the nanowires grown by VLS, the strain could be accurately mapped in the gauge section for seven specimens. For all of them the strain was accommodated homogenously throughout the whole wire. When fracture occurred, almost all wires showed multiple fractures, which is visualized in Figure 5.2. Figure 5.2A shows a specimen during the application of tensile strain shortly before failure. Fracture happened within one frame of the recorded images. Figure 5.2B shows the first frame of a tensile test after fracture. As can be seen the wire still extends from both grips, but the majority of the overall length has disappeared. Many times it was observed that after fracture one side of the wire was still projecting from the grip, whereas the other side failed at or nearby the Pt grip. Figure 5.2C and D highlight the fracture surfaces observed after tensile testing. The fracture plane is always perpendicular to the wire axis. From these images it can also be seen that the surfaces of VLS grown Si nanowires were decorated inhomogeneously to various amounts. This unevenness varied from just a few elevations along the wire to specimens that were fully covered. Also the dimension of these elevations ranges from small particles to secondary Si nanowires that stick to or grow from the tested nanowire.

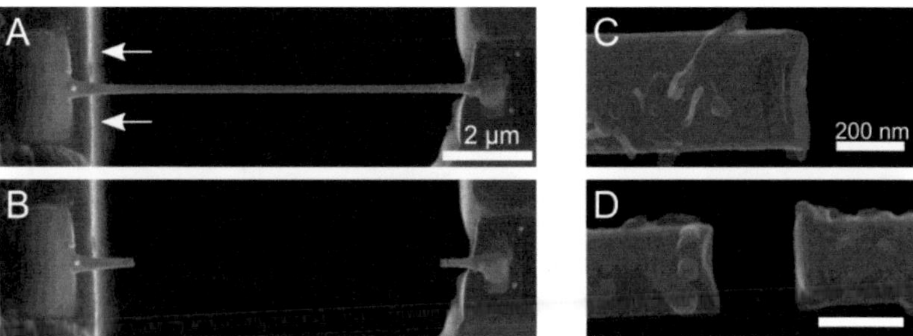

Figure 5.2: Deformation behavior of <111> oriented Si nanowires. The left part of the figure shows a wire during the application of tensile strain short before (A) and after (B) failure. Fracture usually occurs within one frame of recorded images at more than one location and the majority of the overall length disappears. The right part shows two representative SEM micrographs revealing the brittle fracture morphology. Inhomogeneous distortion is visible along the surface.

Since these brittle specimens could not be bent in the same way as the Au nanowires, in order to obtain the actual cross sections the whole SEM stub had to be mounted in a fashion that the wire axes were parallel to the electron beam to facilitate a top-down view on the cross sections of the fractured nanowires. In Figure 5.3 representative SEM images for the tested <111> Si nanowires are shown.

In most cases the cross section was almost circular (Figure 5.3A). However, some wires exhibited cross sectional areas that seemed to be more elliptical (Figure 5.3B). In some cases circles or ellipses that seemed to be truncated (Figure 5.3C) were obtained. Few cases indicated slip at the interfaces between the Pt grip and the wire, as can be seen from Figure 5.3D. As also discussed in A.3.1, these images additionally highlight the trough cutting as a special method to provide thorough gripping of larger specimens. Since these specimens are comparably large and strong, troughs were cut into the substrate before attachment. The wires were then placed into the troughs, so that the Pt could seep around the whole wire creating a strong interface.

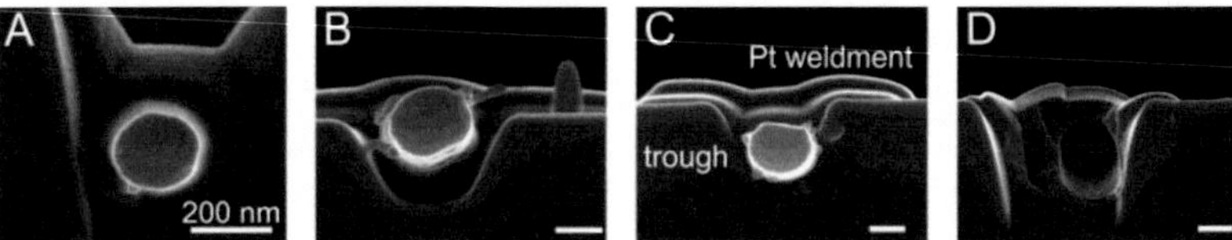


Figure 5.3: Representative cross sections for various wires, obtained by mounting the sample with the wire axes parallel to the electron beam. Circular cross sections are regularly observed (A), but also deviations to elliptical (B) or truncated circles/ellipses (C) can be found. In (D) the wire slipped from the Pt weldment. In all images troughs were cut with the ion beam prior to gripping. The Pt is then able to seep around the whole wire to achieve the strongest bond possible.

The mechanical data for tensile testing of all <111> Si nanowires is shown in Figure 5.4. From both subplots significant scatter can be observed. Figure 5.4A displays the fracture strength as a function of the wire diameter. The class of nanowires grown on SOI approaches smaller scales but shows lower strength and less scatter if compared to the wires grown on $SiO_2$. In Figure 5.4B the Young's modulus measured from the stress-strain curves is plotted against the diameter. Again a lower modulus and less scatter are observed for the nanowires grown on SOI.

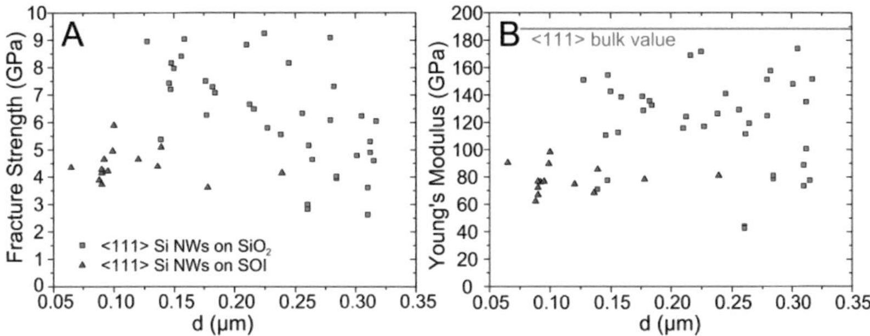


Figure 5.4: Mechanical data of <111> Si nanowires. The fracture strength (A) and Young's modulus (B) are plotted versus the diameter of the nanowires. Whereas large scatter is observed for the one class of nanowires, the other class only shows a narrow distribution for all tests. The measured Young's modulus is significantly below the calculated bulk value.

## 5.2 Tensile Testing of <100> Oriented Nanowires

Silicon nanowires with <100> axial orientation, grown according to the metal assisted chemical etching method described in chapter 1.6, were tested in tension. These wires all had a nominal diameter of about 100 nm and their length varied between 1.5 and 4.5 µm. Among 18 wires tested, all showed a characteristic mechanical response, as shown in Figure 5.5. Rather than start with linear-elastic behavior non-linearity was observed from the beginning. The slope of the stress-strain curve decreased further with continued straining. Although large amounts of strain up to 13% were measured, no necking or pronounced thinning of the specimens was observed in the SEM images. As can be seen in Figure 5.5, all wires failed at high strength of 2.5 to 4 GPa. Fracture occurred abruptly and the fracture surfaces were oriented perpendicular to the wire axis (Figure 5.6C). The slope at the start of tensile testing was around 50 GPa.

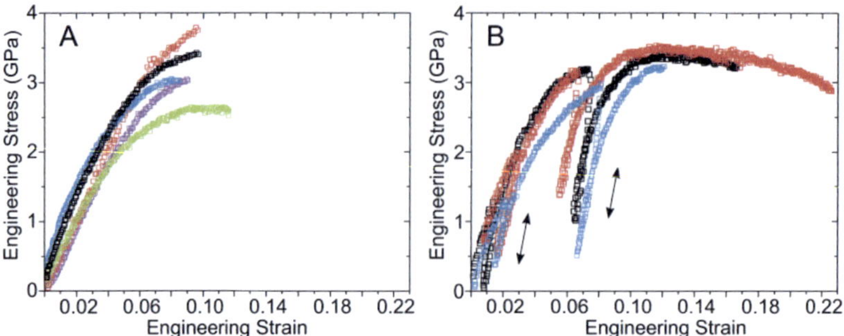

Figure 5.5: Characteristic stress-strain curves for short <100> Si nanowires. The mechanical response shows deviations from linearity from the onset of straining (A). Fracture occurs at high strength and significant amounts of strain. Unloads show a steeper slope (B). Unloaded wires show softening and strains reaching up to 23%.

In a second set of tests, using the same batch of nanowires, unloads were performed to check the elastic-plastic deformation in the specimens. Figure 5.5B shows characteristic stress-strain curves. Like in the previous tests, the modulus at the beginning of loading was around 48±12 GPa and decreased during further straining. However, the unloads showed a much steeper slope, with almost no decrease in strain at the beginning, finally settling at about 120±5 GPa for all tests. The slope again seemed to deviate slightly from linearity. No significant change in the slope could be detected depending on the position of the unload segment in the stress-strain curve. The load segment, which followed the unload deviated from the unloading curve before the initial stress was reached. In these tests the wires sustained even larger amounts of strain reaching up to more than 22%. Also softening in the stress-strain curves was observed (Figure 5.5B). In contrast to the <111> oriented Si nanowires the fracture on multiple positions of the specimen was not observed.

Cross-sectional investigations after fracture revealed an elliptical geometry, as can be seen in Figure 5.6D. The wider side of the ellipse was always facing the electron beam. Markers were placed on the surface of a nanowire and strain was measured for various configurations. Displacement data showed the largest displacement if measured between the two grips. Accuracy could be confirmed for these measurements since the strains in each section (marker-marker, upper grip-marker and lower grip-marker) of the wire summed up correspond to the value measured between the grips. This is different if the displacement data is converted into strain. The strain data (according to the colors) and an SEM micrograph are displayed in Figure 5.6A and B. It can be seen that, in contrast to the displacement data, the largest strain measured by DIC is returned in the gage section between the two markers. At the transition of the gage section (markers) to the grip the strain was significantly lower.

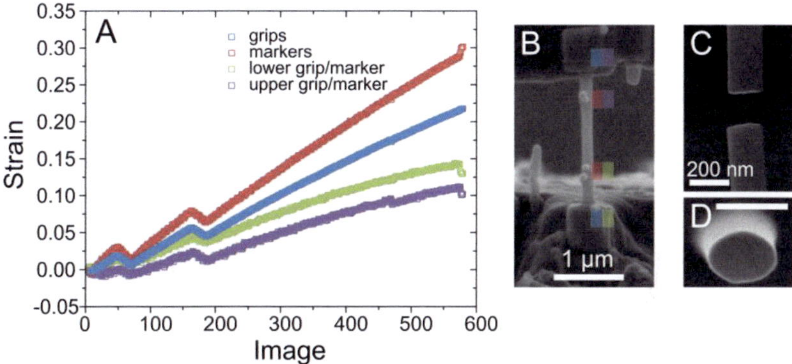


Figure 5.6: Strain versus image number for DIC performed on various positions of a nanowire (A), corresponding to the respective color in the SEM image (B). Fracture surfaces of the nanowires were always perpendicular to the wire axis (C). The cross sections obtained after fracture are elliptical (D).

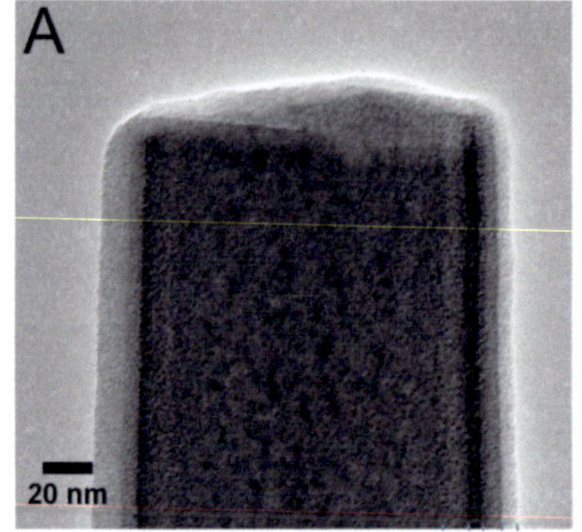

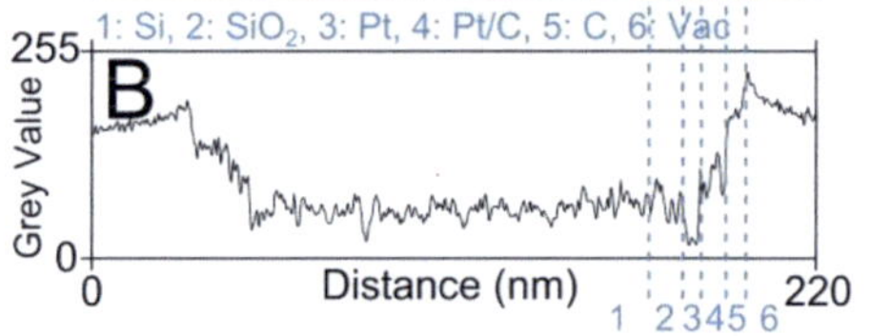

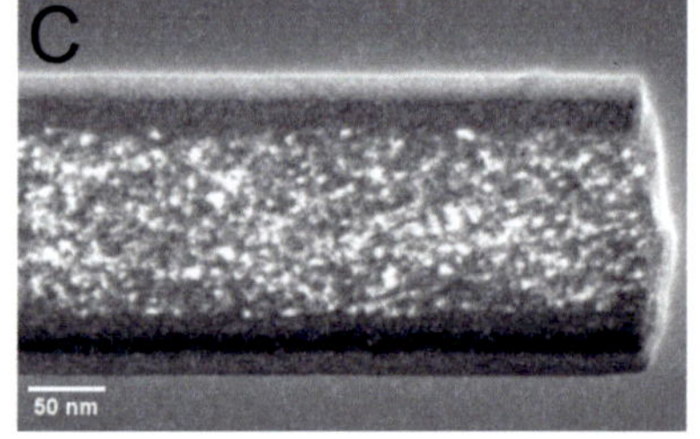

Figure 5.7: TEM image of tensile tested short Si <100> nanowire (A). The micrograph shows contrast variations across the width indicating different materials. If the grey values are taken (B) the number and width of the single layers is revealed. Superposition of a bright field and a dark field TEM image (C) show the crystalline core of the nanowire.

TEM analysis was performed on eight fractured portions of the strained wires (Figure 5.7). Strong surface decoration containing heavy elements, which most likely derived from Pt-organic deposition, impeded the acquisition of clear images of crystalline Si (Figure 5.7A). To circumvent this, a dark field and a bright field image with the same magnification were taken and superimposed. In this configuration the crystalline material returned a bright contrast whereas other regions appeared dark. An exemplary superposition can be obtained from Figure 5.7C, where the heavy Pt on the surface caused a speckled pattern. Figure 5.7A shows the layered structure of the nanowire. The crystalline Si core is approximately 100 nm thick and the outer shell can be divided into four parts. The Si core is followed by a layer of native oxide on the order of 15 nm thick. On top of this layer a 10-15 nm thick layer of Pt-deposit has formed. From zoomed out TEM images it is observed that a conformal Pt-organic layer spreads along the entire wire. Following this Pt-organic layer is a mixture of Pt-organic composite and carbonaceous contamination, which then results in the outermost layer that solely consists of carbonaceous contamination deposited during e-beam imaging. These three layers could also be regarded as one layer with a gradual change in Pt content. The fracture surfaces themselves are contaminated and no sign of plastic deformation is visible.

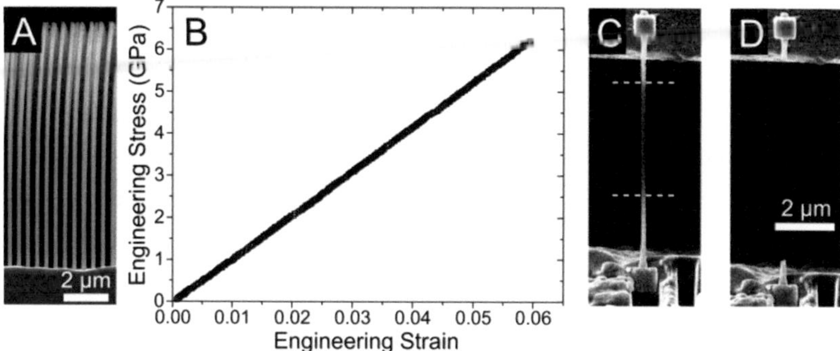

Figure 5.8: Long Si <100> nanowire specimens (A) (l ≥ 8µm) show linear elastic loading up to high strength (B) with brittle fracture at multiple locations (D). A strong change in contrast is observed along the wire axis (C).

In contrast to the short <100> nanowires (≤ 4µm) longer specimens (≥ 8µm, see Figure 5.8A) showed different behavior. All five tested specimens showed high strength between 4.0 and 6.4 GPa and Young's moduli of 84 to 105 GPa. Figure 5.8B shows an exemplary tensile curve. Deformation was solely elastic until brittle fracture occurred. In contrast to the short specimens fracture on multiple locations took place (Figure 5.8D), as previously described for the <111> oriented wires (cf. chapter 5.1). In Figure 5.8C an SEM micrograph of the initial tensile configuration is shown with modified contrast. It can be seen that the outer regions of the specimen are brighter than the middle section of the wire, indicated by the dashed lines, which is indicative of a strong coating close to the gripping points. Subsequently 2 wires were coated with Pt-organic composite to produce the same conditions as for the short specimens. No change in the mechanical response could be observed and the wires failed at similar stresses (4.1-5.9 GPa).

## 5.3   Discussion

The <111> oriented Si nanowires show large elastic strains and high strengths approaching the lower calculated limit of the ideal strength in this material [18]. Fracture is always brittle and no hints of plastic deformation could be observed. Large scatter is obtained in the mechanical data of fracture strength and Young's modulus. <100> oriented nanowires showed distinct mechanical behavior. Also large stresses and strains were reached. The long specimens showed behavior similar to the <111> oriented nanowires, exhibiting linear-elastic loading until brittle fracture. For short specimens the tensile curves showed characteristic nonlinearity from the start of straining. In contrast to this observation, deformation morphology showed brittle fracture and no strong evidence of plastic deformation.

Fracture always occurred abruptly in <111> Si nanowires and on more than one location within a single frame, too fast for the image acquisition of the SEM to catch it. In most cases the initial fracture point and the initial fracture surfaces could not be observed. This type of fracture is known as "Spagetthi fracture" [165]. A tremendous amount of elastic energy is stored in the deformed wire and is suddenly released upon fracture. Waves run through the sample, reflect and superimpose with other waves to form stress concentrations at which the wires will break again. A second possibility for multiple fracture could be the feedback loop of the transducer, which at fracture causes a forward surge of the transducer tip and then comes back to meet its regular position, as described in Appendix A.1.3. During the return the transducer may overshoot again and cause contact and fracture of the wire portions. However, this alternative can be ruled out since this type of fracture is independent of the employed testing setup (chapters 2.1.2 and 0).

The Young's moduli of <111> nanowires grown on $SiO_2$ show large scatter (Figure 5.4B). Since no trend is discernible and no good physical reason would hold for the observed scatter, there are various parameters, which could contribute to the observed deviations. The following four parameters will be briefly discussed: (i) cross-sectional measurement, (ii) oxide-layer, (iii) misalignment and (iv) Pt compliance. Measurement of the cross sections is straightforward for these brittle specimens. The stub is mounted with the tensile axis parallel to the e-beam and the cross section can be directly measured, as depicted in Figure 5.3. Large uncertainties cannot be expected from this parameter. Also a contamination layer as observed for the Au nanowires is not observed to this extent. Smaller contamination layers may be existent, but would not have a strong effect on the effective modulus since it is more than double that of Au. This is different with the oxide layer, which is native for silicon and therefore exists on all of the nanowires. The oxide layer of Si is usually in the range of a few nanometers. The Young's modulus of this oxide layer is reported from 45 to 70 GPa having a much more pronounced effect than a contamination layer. The oxide layer is appearing similar to pure Si and therefore not distinguishable in secondary electron (SE) images during cross-sectional measurements further complicating the interpretation. Using the approach of the Voigt composite model, as discussed in chapter 2.2.3, the volume fractions of the oxide layer were calculated. Figure 5.9B shows the calculated oxide layer thickness in dependence of the measured wire diameter having a mean value of 23±17 nm. This value is very high if compared to the standard native oxide thickness on Si (1-3 nm), but can still be possible due to the different method of synthesis for these nanostructures if compared to bulk Si. Growth of nanowires by VLS employs Au as a catalyst, which enhances oxide formation in air at RT. Also the thickness of the oxide layer may vary depending on different amounts of Au in different wires. However, for small moduli the volume fraction would exceed 100% in these calculations, which is impossible. Although the oxide layer is definitely affecting the mechanical response it cannot be the only reason for the observed scatter. Another error source could derive from misalignment. The specimens are rather short and therefore misalignment has a larger effect on the measurements than in long wires. Misaligned nanowires appear shorter in the 2D SEM images and as a consequence the strain is overestimated, which finally leads to a lower value for Young's modulus. An indication of misalignment is given by the location of fracture. In a brittle material with a constant cross section fracture can basically occur anywhere in the gage section. However, fracture at the grip is indicative of stress concentrations due to superimposed bending stresses through nonuniaxial testing. The amount of misalignment could be examined during cross-sectional analysis (Figure 5.3), where the deviation of the nanowire axis from the tensile direction could be determined by tilting the sample. Misalignment of the specimen in the focal plane of the microscope causes a change in the observed length of the specimen, following a cosine relationship. This only gives small deviations for small misalignment, but increases rapidly for larger misalignment. For all four nanowires observed with a misalignment >10°, the measured Young's modulus was below 75 GPa, validating a strong error for heavy misalignment. The forth reason for the observed scatter may be the compliance of the Pt grips. Although this did not seem to be a problem in the Au nanowires, the Si nanowires exhibit much higher stresses that need to be sustained by the grips. Furthermore, the diameters of these nanowires are larger, which makes the achievement of a strong bond much more difficult, as mentioned in chapter A.3.1. As a hint of compliance of the grips the Young's modulus should scale with the length of the nanowires. However, no strong correlation was found (Figure

5.9A) for these specimens. This may be due to variations in the Pt deposition. Although consistent beam settings were used for deposition, the area and length of deposition as well as the surfaces of the nanowires and the substrate were always different causing each deposition to be unique. While strain measurements on the Pt grips did not reveal deformation, motion of the grips could occasionally be observed in the image right after fracture. In few cases the wires slipped at the grips or at the deposit interface as shown in Figure 5.3D. Overall the stochastic nature of the elastic properties measurements may be caused by a combination of oxide layer thickness, misalignment and grip compliance.

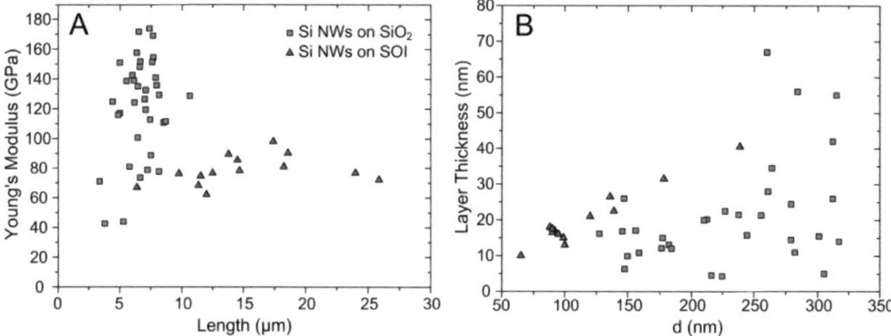


Figure 5.9: Length dependence of Young's modulus of <111> Si nanowires (A) shows large scatter for one class, whereas a slightly increasing modulus and only small scatter is observed for the other class. The calculated layer thickness (B) shows significant scatter with a mean value of 23 nm for one class. In second class the layer thickness increases as the diameter of the specimen increases.

Measurements of the Young's moduli of <111> oriented Si nanowires grown on SOI did not scatter to the extent as for the nanowires grown on $SiO_2$. However, all measured values were significantly below the theoretical values for bulk <111> Si. Except for the two longest specimens in Figure 5.9A, a trend for an increase of the measured modulus with increasing length can be observed, supporting the argument of compliance of the Pt grips. In contrast to the nanowires grown on $SiO_2$, the calculated oxide layer thickness increases with increasing wire diameter (Figure 5.9B) and thus part of the observed differences may be attributed to the different growth parameters. First and foremost, the data of Figure 5.9A in conjunction with SEM investigations suggest that the different amounts of scatter may mainly be caused by different amounts of misalignment of the specimens in the focal plane. On the one hand the nanowires grown on $SiO_2$ grow out of a Si substrate and may be inherently misaligned, whereas the specimens grown on SOI lie flat on the substrate, from which they are harvested. This substrate should be in good alignment with the focal plane and therefore no strong misalignment can be expected from this batch. On the other hand small deviations in z-direction of the two bearings have a much more pronounced effect on short specimens, which is further substantiating the argument.

In the context of mechanical size effects the fracture strength of Si nanowires of both growth methods is compared to the results of other small scale mechanical experiments (Figure 5.10) as nanowire tension [130, 166], nanowire bending [167], microwire tension [125] and micropillar compression [168]. Comparison of the different results has to be done carefully, since the acquisition of data has been achieved using distinct approaches with varying accuracy. While Zhu *et al.* [130] simply measured the deflection of an AFM cantilever to calculate a stress-strain curve using the known Young's modulus for bulk Si, Steighner *et al.* [166] measured the real force on the specimen. In their microwhisker tensile experiments Cook *et al.* calculated the diameter of their specimens from the slope of the stress-strain curve and the known values of Young's modulus of bulk Si [125]. Bending experiments, as performed by Hoffmann *et al.* [167], exhibit non-uniform stresses that complicate interpretation. The data from this work can be used in its raw form (Figure 5.10A) or it can be treated to account for uncertainties. The corrections for the Si nanowires can be done in two different ways. Assuming a non-negligible oxide shell, the Voigt model, as described in chapter 2.2.3, can be employed using the known Young's modulus of Si and $SiO_2$ to calculate the true thickness of the wire. In the second case, assuming uncertainties in cross-sectional measurements, the diameter of the specimen is adjusted, so that the Young's modulus matches the expected value. Figure 5.10B shows the data of the Si nanowires corrected for the latter option. From Figure 5.10 it can be seen that the data of this work lie between the ideal strength of <111> pure Si [18] and the results obtained for tensile testing of thin films [169], which had a nominal thickness of 2 µm. Especially for the corrected values (Figure 5.10B) a trend is visible suggesting a steady increase in strength up to a diameter of ~200 nm followed by a flat regime of fracture stresses close to the lower theoretical limit of the strength.

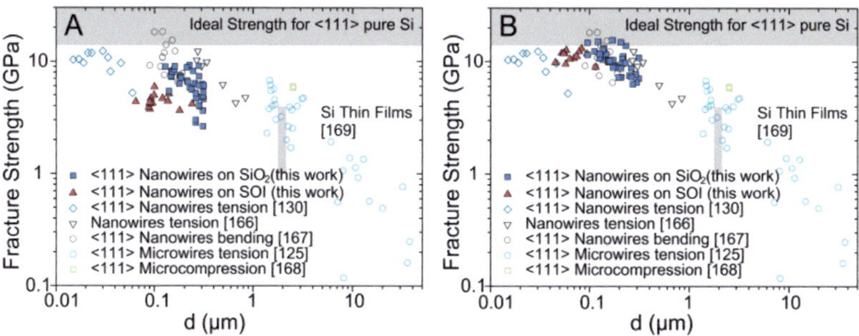


Figure 5.10: Masterplot of Si small scale mechanical measurements for measured (A) values and corrected for Young's modulus (B). The results of tensile testing of nanowires are compared to other small scale mechanical experiments of Si.

As already mentioned for the Vanadium dioxide nanowires, the mechanical behavior of brittle materials can be described by weakest link statistics, where the probability of finding a critical defect determines the strength. A Weibull analysis was performed on both groups of nanowires. Since there was no large population having similar volumes or surface areas, as required for standard Weibull statistics, a different approach, shown in Appendix A.4, had to be employed to obtain reliable data. The calculated Weibull moduli are given in table A.1. Figure 5.11 shows Weibull plots for fracture strengths versus volume, surface area and length for both groups of nanowires, as well as for the measured and corrected data. Although a faint increase in fracture strength with decreasing sample dimensions can be observed no clear correlation is visible. In contrast to the expectations from Figure 5.10, the Weibull moduli for the wires grown on $SiO_2$ do not change significantly between the data for measured ($m_{surf} = 4.3$, $m_{vol} = 4.5$) and corrected ($m_{surf} = 4.8$, $m_{vol} = 5.0$) results. Also, these moduli are at the lower end of typical moduli expected for ceramics (5-20) in line with the lower values of Weibull moduli obtained for thin film tensile testing [169]. In contrast, nanowires grown on SOI exhibit a modulus on the upper limit of values ($m_{surf} = 7.2$, $m_{vol} = 7.8$) reported in thin films, which is further increased if the corrected data ($m_{surf} = 11.5$, $m_{vol} = 11.8$) are used.

For a pure material in a defect free state, fracture is expected to occur at the level of the ideal strength without large variability. In contrast to these assumptions the data for Si nanowires grown on $SiO_2$ exhibit larger scatter and fracture stresses which are below the lower theoretical limit for both (measured and calculated) groups. Furthermore, two batches of different growth methods with the same axial orientations in the same size regime show different behavior. Whereas the data for one group is largely scattered the data of the other group is more or less constant.

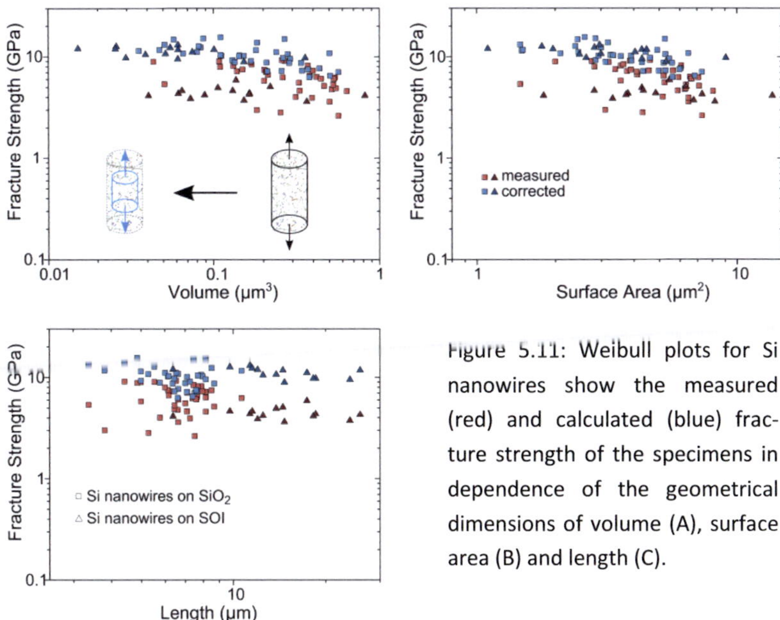



Figure 5.11: Weibull plots for Si nanowires show the measured (red) and calculated (blue) fracture strength of the specimens in dependence of the geometrical dimensions of volume (A), surface area (B) and length (C).

The calculated Weibull moduli revealed that the variability in the data for the nanowires grown on $SiO_2$ does not depend on any corrections, but is inherently existent in this material. In the other group the corrections are further raising the modulus, approaching values that are in the upper regime for the Weibull modulus of brittle materials. Explanations for this distinct behavior can be drawn from various points. The difference in variability in the recorded fracture strengths points towards a different number and character of defects or flaws in the material, implementing two possibilities: (i) volume defects or (ii) surface flaws. Defects in the volume of the specimens may most likely be foreign Au atoms, which are accumulated from the precursor during growth. Since this is a highly stochastic process strongly depending on the local conditions these atoms may also cluster to form larger impurities, thus further lowering the fracture strength of the material. Due to the difference in growth, more Au is expected in the specimens grown on $SiO_2$ [170], which is supporting the findings of a larger scatter in these specimens. Also the lower fracture strength in both groups could be attributed to the existence of these volume defects. However, the most striking argument for the obtained difference in Weibull moduli are the different amounts of irregularities found on the surfaces. Whereas the wires grown on SOI show consistently clean surfaces, the specimens grown on $SiO_2$ exhibit various amounts of surface contamination, ranging from small dots to branch wires extending from or wrapping around the parent wire. These contaminations provide notches, which in turn serve as starting points for crack initiation and subsequent failure. Bryans *et al.* showed that the strength of Si whiskers could be improved by decreasing the size of surface steps [171]. Also the density of surface decoration is found to strongly vary. To investigate this effect, the specimens were classified depending on the amount of the surface irregularities. Although a stronger contamination should result in a higher probability of finding a critical flaw, no correlation between the fracture strength and the amount of irregularities was found. It is possible that more statistics is needed to reveal such dependence.

A different behavior is observed for two batches of Si nanowires with <100> axial orientation. For the short <100> oriented nanowires manipulation is impeded since harvesting and gripping additionally shorten the specimen. For these short wires, never exceeding a length of 4 µm, the contamination of surfaces is stronger than in the specimens investigated so far. Decoration of surfaces is seen in TEM images, where the heavy Pt, existent from EBID, inhibits the observation of crystalline silicon. The superposition technique shown in Figure 5.7 allows for exact measurement of the Si core, which is about 100 nm in every wire. This is in good agreement with the predictions of the synthesis where the mask was tuned to produce nanowire arrays with diameters of ~100 nm. The Si core is covered by an oxide shell, larger than the commonly observed native oxide, which is reasonable since the wires are synthesized by a wet chemical process in which the oxide layer is not natively grown by the exposure to ambient air, but during the chemical process. Two dark lines follow the oxide layer, which belong to the two Pt depositions for gripping. The deposit is diffusing along the wire length forming an amorphous coating. On top of this layer a bright layer is visible originating from e-beam imaging and the deposition of carbonaceous material from the chamber. The long <100> specimens further substantiate the findings of a strong contamination. In Figure 5.8C a strong change in contrast is visible along the length of the specimen. The bright contrast close to both grips first emerged after the deposition of the Pt-organic compound. From the images it can be seen that the Pt diffuses along the nanowire surface (bright area), leaving a conformal coating. The diffusion length of the deposit at the employed beam settings ( 5 kV, 0.41 nA) is determined to be ~2 µm, which is the distance from the edge of the deposited grip to the point at which the contrast fades.

To investigate the influence of the Pt-organic coating on the mechanical behavior of the nanowires two long specimens were intentionally coated with Pt in two different ways; (i) imaging right after the grip deposition with closed valve but the precursor gas still in the chamber and (ii) deposition in the gauge section with opened precursor valve, where the latter causes a larger layer. Subsequent tensile testing showed the same behavior as for the uncoated long <100> nanowires. Therefore the Pt layer can be excluded from governing the change in mechanical response.

With the effect of Pt contamination disqualified there are three other possibilities that could influence the mechanical behavior. (i) corrugations along the nanowire surface, (ii) doping or (iii) porosity of the crystalline core. In the SEM images corrugations are frequently observed along the length of the specimens. Since the etching process is very accurate these corrugations are most likely generated by errors in the metal etching mask. However it is not very likely that corrugations parallel to the tensile direction affect the mechanical behavior. Differences in doping exist between the two batches of Si <100> nanowires. The short specimens are n-doped (0.005 $\Omega$cm), while the long wires are p-doped (1-20 $\Omega$cm). These differences may result in different concentrations of vacancies in the lattice that could lead to the marked changes in behavior. Further it is known that high doping can cause porosity [170], which would significantly change the mechanical response. The correct modulus measured during unloading suggests that there is at least a fraction of <100> oriented Si in these specimens. If the wire is unloaded during testing the fracture strain is considerably increased, pointing towards an influence of relaxation. Strain measurements do not indicate an influence of the grips on the recorded curves, leaving differences in doping as the only possibility for this distinct behavior. From this point of view no clear conclusion can be drawn and the question remains unresolved. More investigations of the microstructure are needed to reveal the origin of this marked change in behavior. It is evident that a size effect on BTD transition, as shown by Östlund *et al.* [12], can be excluded since both batches of wires have similar diameters, which are significantly smaller than the mentioned transition region.

# 6  Sodium Chloride Nanowires

NaCl is probably the most commonly known ionic structure, owing to the fact that mankind has to deal with it in everyday life, e.g. as table salt. From this experience it is also common knowledge that sodium chloride is inherently brittle and is therefore expected to cleave upon application of stress, although it was already observed back in the early 1920s that rocksalt can be ductile. The rocksalt structure exists for configurations of ionic radii below 0.732, where the coordination number of the cation decreases to six. The structure is fcc for each type of ions and can be described by two interpenetrating lattices. Much work was performed on NaCl in the $20^{th}$ century due to the relative ease of preparation and the visible formation of slip traces on the surface facilitating the observation of plastic deformation.

## 6.1  Tensile Testing of Sodium Chloride Nanowires

NaCl nanowires with widths ranging from 125 to 560 nm were tested in tension in 18 individual experiments. Experimental parameters were used according to the settings mentioned in chapter 3 and chapter 4. Despite two cases, all wires showed elastic-plastic transitions at yield stresses between 120 and 250 MPa. Figure 6.1A shows characteristic stress-strain curves of various specimens. Linear-elastic loading was always followed by plasticity at a constant flow stress of the level of yield. In some cases faint degrees of hardening could be observed. To confirm the plastic nature of the plateau some specimens were elastically unloaded during the experiment, as exemplified in Figure 6.1B. Fracture occurred at total strains of 2 to 14 %. In most cases no hardening was observed.

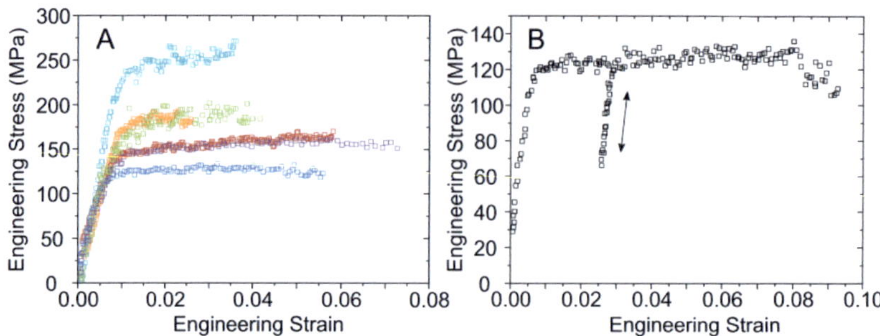

Figure 6.1: Characteristic stress-strain curves for various NaCl nanowires (A) show elastic loading followed by strong plastic deformation at the level of yield up to total strains of 14 %. Unloading (B) during the experiment confirmed the plastic nature of deformation.

The onset of plastic deformation in the curves could be correlated to thinning and subsequent formation of necks in the SEM images (Figure 6.2). Although hardening is observed for this particular specimen the deformation behavior is similar to specimens that do not show hardening in their curves. Upon yielding extended plasticity was visible. Plastic deformation localized and sometimes multiple necks formed. During further straining the necks and subsequently significant parts of the gage section thinned by varying amounts until fracture occurred abruptly. However, the wires did not necessarily fail at their thinnest section (see IV, Figure 6.2B). The fracture surfaces of the specimens were always oriented perpendicular to the tensile direction (Figure 6.2B).

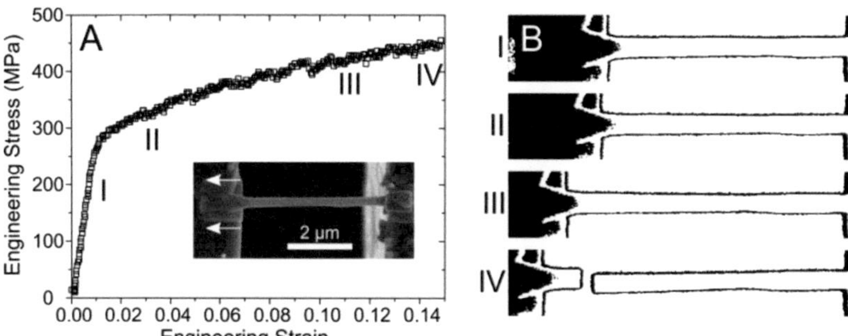

Figure 6.2: Plastic deformation in NaCl nanowires. The mechanical response (A) is correlated to the deformation behavior (B). During linear-elastic loading the wire is strained homogeneously (I). Upon yielding, faint thinning along the specimen is visible (II). At large displacements (III) the wire has thinned significantly and necking is more pronounced in the region indicated by the dotted circle. Fracture does not occur in the thinnest region (IV). The fracture surfaces are perpendicular to the tensile direction indication brittle fracture.

In some cases the fractured portions of the wires could be regripped and pulled again. Figure 6.3 shows three consecutive tests on the same specimen. Little plastic strain was observed for the first test. Yielding of the second test roughly started at the level where fracture had occurred before. This test showed significant strain hardening and a high degree of ductility during deformation to a total strain of 15%. For the third test yielding started slightly below the level of precedent fracture stress and showed a still higher strain hardening rate as before. The stress at failure was more than twice as the initial yield stress of the first test.

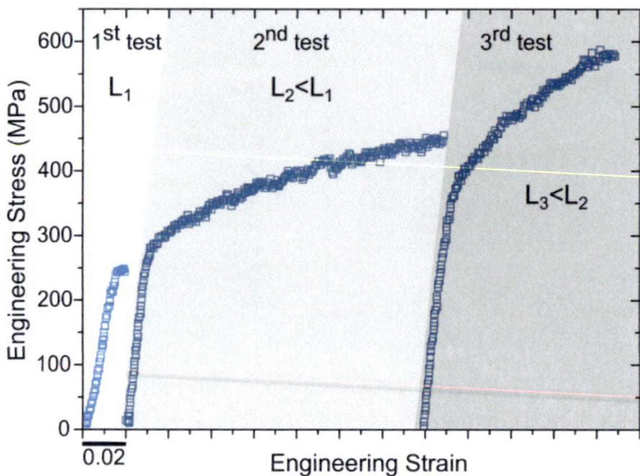


Figure 6.3: Three consecutive tensile tests on one individual NaCl nanowire. After little plastic deformation in the first test yielding starts at the level where fracture occurred before. Significant strain hardening and plastic deformation to large amounts are observed in the second and third test, where the final fracture stress is more than twice as high as the initial yield stress. Since one of the fractured portions is regripped after fracture the total length of the specimen decreased with every test.

The cross sections were examined after testing, following the approach of chapter 5.1. The wires exhibited cross sections, which were mostly square with rounded edges and in some cases rectangular with the highest aspect ratio being 2:1. Representative images can be seen in Figure 6.4. EBSD investigations confirmed that the wires were single crystalline and oriented with their tensile axes along <100> directions. No change in orientation was observed throughout the wires, also in the region of fracture. Voltages of 20 kV at beam currents of 2.1 nA were used for EBSD measurements in order to obtain strong enough signals for these small specimens. When the specimen was irradiated, the corresponding Kikuchi pattern was vanishing after a short time. The nanowire showed visible traces of the electron beam on its surface, ranging from charged spots to altered regions. Tensile testing was performed at lower voltages, typically 5 kV, and lower beam currents of 98 pA to not affect the structure of the nanowire. EDX analysis showed an ion ratio of 50:50, as expected for pure NaCl. Since the maximum net exposure time to the e-beam in one single test is about 30 minutes, NaCl crystals were imaged for the same time and EDX spectra were acquired every 90 seconds. Figure 6.4 shows a stable composition during the whole experiment. However, a decoration of the surface became visible after a couple of minutes, which is further increasing with imaging time.

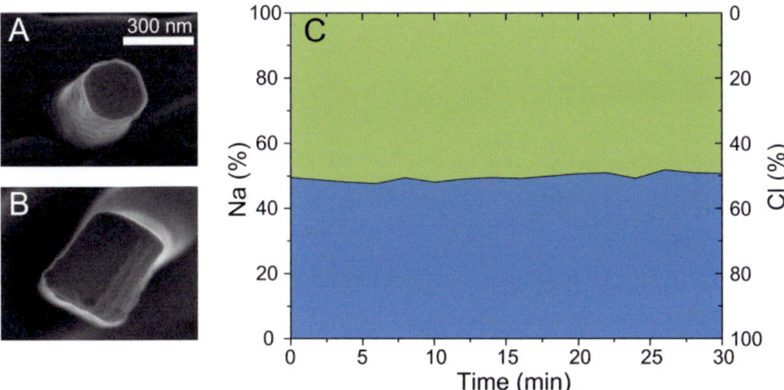

Figure 6.4: Representative cross sections of NaCl nanowires. Most specimens exhibited square cross-sectional areas with truncated or rounded edges (A). In some cases rectangular cross sections were observed (B). Imaging a NaCl surface for 30 min at the beam settings of tensile testing and concurrently acquiring EDX spectra showed stable stoichiometry of NaCl (C).

The observed yield strength varied between 120 and 250 MPa. Figure 6.5 shows the mechanical data of yield strength and Young's modulus plotted versus the nominal edge length of the cross section. From Figure 6.5B it can be seen that the modulus is only half of the value for bulk NaCl calculated from the elastic constants.

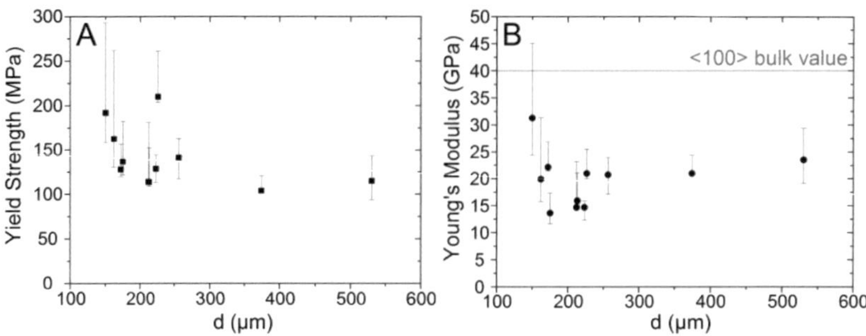

Figure 6.5: Yield strength (A) and Young's modulus (B) plotted versus the nominal edge length. The Young's modulus is only half of the calculated value for bulk NaCl.

## 6.2 Discussion

Tensile testing of NaCl nanowires reveals strong plastic deformation at stresses higher than observed before but still significantly below the theoretical strength, for which the lowest calculated value is 2.39 GPa [172]. Plasticity manifests by elongation and thinning of tremendous parts of the gage section sometimes accompanied by local necking. The elastic-plastic deformation behavior was confirmed by unloading and reloading a specimen in the plastic regime (Figure 6.1B). However, fracture did not always occur at the thinnest section of the wire and the fracture surface was oriented perpendicular to the nanowire axis and was showing a mirror clean surface, indicating brittle fracture.

The EBSD measurements showed a consistent pattern throughout the nanowire confirming the single crystalline microstructure. The axis of the investigated wire was oriented along [001] and did not change, even at the site of fracture. Therefore a change in plastic deformation mechanism is not expected. The square or rectangular cross sections of the specimens with side surfaces oriented perpendicular to each other point towards {100} or {110} orientation of the side surfaces, where the surface energies favor

{100} planes. It is possible that additionally {110} surfaces truncate the edges of the {100} side surfaces or vice versa, which would lead to the observed cross-sectional geometry in Figure 6.4A.

Ductility in Sodium Chloride has already been reported in the early 20th century [52] and is commonly known to appear for other ionic materials as well [49]. Recently the formation of salt nanowires and their superplastic elongation have been observed in TEM investigations [173]. Nevertheless it was also reported that the composition of Sodium and Chloride ions was dramatically changing, due to the heavy irradiation with the electron beam. It is known that the exposure of NaCl surfaces to electron irradiation introduces defects, which can rapidly migrate to surfaces and further can cause the formation of nanocrystals [174]. However, decomposition of the material is the most concerning fact. Upon illuminating NaCl with high energies inside the TEM NaCl decomposes, the chloride being volatile, and an increase in sodium content is observed. Imaging a forest of nanowires is possible for about 20 seconds before the stoichiometry is changed drastically [175]. The effect of the electrons on the structure of NaCl strongly depends on the settings, which are much lower for the tensile experiments inside the dual-beam microscope if compared to TEM. Continuous EDX measurements for 30 minutes, which is an upper limit of the net time of exposure during such an experiment, were performed to check for possible decomposition under experimental settings of 5kV and 98pA. As shown in Figure 6.4C, no change in stoichiometry could be observed and therefore no effect on the tensile experiments is expected. However, an increasing surface distortion leading to the formation of nanocrystals could be seen, but was never observed during the tensile experiments.

Plastic deformation to large amounts of strain was observed in 10 out of 11 wires. In a material exhibiting both, cleavage and glide planes, the cleavage is usually favored in bulk materials, since defects and flaws will lower the stresses needed for crack initiation and propagation leading to subsequent failure by separation of two layers of atoms/ions. In these small specimens the ductile behavior in the nanowires may on one hand be owing to the high crystal purity of the material, which does not provide easy sources for cleavage, and on the other hand the smaller size holds a lower risk of finding a critical defect. That being said, the CRSS favoring glide on {110} <110> (m=0.5) may be larger than the critical resolved normal stress (CRNS) favoring cleavage, because these pure crystals with high surface quality do provide no/less defects, which would increase the CRNS. The same is valid for the influence of humidity. Although the wires were mostly kept in evacuated environments, trace amounts of water can already have an effect. In early publications it is hypothesized that water heals out defects on the surfaces of NaCl crystals, which would lead to the same effect as described for the initially pure material. As increasing temperature is known to enhance ductility, the input of heat through electron bombardment may also lower the critical stress needed to commence glide. However, the heat produced through e-beam irradiation should be negligible. For the <100> oriented wires four equally preferred {110} <110> primary glide systems are available. From the SEM images it is not possible to tell whether all of them are active. The thinning and the concurrent mechanical response, which is almost ideal plastic, suggest deformation by the sequential nucleation and multiplication of dislocations on primary glide systems. Varying amounts of hardening in the stress-strain curves point towards easy glide and a constant nucleation stress for no hardening and on the other side interaction of dislocations with each other or obstacles, leading to increasing nucleation stresses in the wires showing hardening. Obstacles as vacancies or interstitials are known to charge dislocations, which can further impede their motion [49]. Charging was also induced by the electron beam and therefore might affect the mechanical response.

If the shape of the recorded tensile curves is compared to that of the characteristic stress-strain curves for compression of single crystals [176] at experimental strain-rates between $10^{-3}$/s and $10^{-5}$/s, similarities are obvious. The shape of the tensile curves corresponds to stage I in compression, which as well is characterized by easy glide and low and constant hardening rates. Whereas compressive deformation continues with stage II, which is a second linear hardening region with a larger rate and stage III with a decreasing hardening-rate, fracture occurs for the tensile specimens. In one case (Figure 6.3) a similar curve could be obtained for consecutive loading of a nanowire, which showed different amounts of hardening. The reason why tensile stress-strain curves only show stage I behavior may lie in the different effect of compressive and tensile stresses on crack growth, which will be taken up again later. The statistical nature of events can be seen in the stress-strain curves of Figure A.5.1, where fractured NaCl nanowires were consecutively tested. The curve fits show that plasticity itself may not be the reason for fracture, which coincides with the observation of global deformation throughout the wires terminated by brittle failure.

Fracture always occurred abruptly and the fracture surfaces were oriented perpendicularly to the tensile direction, as observed in brittle fracture. Also the cross-sectional investigations (Figure 6.4A and B) show a clean mirror surface, which is indicative of brittle failure. Further, fracture not always took place at the smallest width, i.e. the smallest cross section, indicating an event at a different site with larger cross section. There are several possible explanations for this behavior. The first is embrittlement caused by electron irradiation. Also after relatively heavy irradiation ductile deformation behavior is observed [173] excluding this mechanism from the map. Second, the fracture takes place on a {100} plane, which is part of the secondary glide system in NaCl crystals and therefore the failure could be caused by activation of secondary glide. However, according to Schmid's law the CRSS perpendicular to the tensile direction is zero and glide on this system is not expected. Only misalignment or crystal rotation during testing could cause activation. Whereas the former was checked and can be ruled out, the latter is known to happen due to grip effects [49], in which the glide direction rotates and secondary glide systems are activated. Nevertheless at fracture, this should still lead to a fracture surface that is inclined to the tensile direction, which was never observed. Possible reasons for this change in deformation are (i) the creation of surface steps, (ii) the loss of mobile dislocations or (iii) the formation of nanocrystals on the wire surface due to irradiation, which act as flaws. In a material that exhibits both cleavage planes and glide planes, small changes may lead to the activation of a different mechanism of deformation. The first case implies the formation of slip steps on the nanowire surface during easy glide. They subsequently lower the activation stress for cleavage, which is then suddenly preferred and results in the observed brittle fracture. Cleavage does not necessarily have to happen at the smallest wire cross section, since local stress concentrations at the created surface steps will initiate cracks independent of the wire dimensions. In bulk NaCl the emission of dislocations from a crack tip at very small velocities of crack growth could be observed. Here the inverse case is observed, namely the nucleation and storage (multiplication) of dislocations and subsequently the nucleation and propagation of a crack. In the second case, constant nucleation of dislocations causes steady deformation of the crystal. Although the material is single crystalline the nucleation is relatively easy at the beginning. During the course of plastic flow dislocations interact and hardening occurs, which is mostly invisible in the engineering stress-strain curve, but could be observed if true stress-strain data is plotted. Plastic deformation continues until no more mobile dislocations can be generated or exist, i.e. the material would simply run out of mobile dislocations. As a consequence, plastic deformation is terminated in this region. Brittle failure suggests that the cleavage is triggered by a flaw on a different site that causes a crack and subsequent failure. Using a simple criterion for a penny-shaped crack in an infinite medium the critical flaw size can be calculated. With the fracture stress $\sigma$

and the stress intensity factors $K_{Ic}$ obtained from literature, ranging from 0.4 to 0.75, the critical crack length is between 3µm and 8µm, which is more than an order of magnitude higher than the actual width of the specimen. This approach may not be sufficient, since a specimen of these dimensions cannot be regarded as infinite. If a weight function is applied to account for the dimensions the value for the critical flaw size is reduced, however it is still exceeding the width of the specimen by far. Therefore the question arises, whether a continuum mechanics approach can be used for this question on the atomic scale. It is also possible that there is no conventional crack and the assumption of a plastic zone is more appropriate here. Another question is the validity of the measured $K_{Ic}$ values in these small dimensions. In the third case, the formation of nanocrystals on the wire surfaces, as observed for the NaCl surfaces in EDX, would lead to flaws and stress-concentrations, basically following the explanations of the first case. The occurrence of brittle fracture by cleavage explains the observed differences between tension and compression in the previous paragraph. Since compressive stresses would not enforce crack-growth, plastic deformation continues but has to overcome higher stress barriers due to interaction of dislocations. In tension crack growth is possible and hence the mechanical response is terminated with brittle fracture after stage I.

Figure 6.6 shows the yield strength of the nanowires plotted versus their edge lengths for the measured (Figure 6.6A) and corrected values (Figure 6.6B). The correction was made similar to the Si nanowires, where the cross section of the specimen was adjusted for the expected Young's modulus, assuming uncertainties in the cross-sectional measurements. Furthermore, the results for tensile experiments on microwhiskers are included [135]. The calculated ideal strength for NaCl has a pronounced minimum in <100> which is increasing towards <110> and <111>. In <100> directions the calculated values span a large range from 2.4 (constrained) to 25 (unconstrained) GPa in tensile and 2.6 to 12 GPa in shear direction, respectively [18]. It can be seen that the obtained values strongly exceed the reported values for single crystal compression [51, 61], which are in the low MPa regime. However, the values of the calculated theoretical strength are still an order of magnitude larger. If compared to the data obtained by Gyulai [135] the values are lower than the highest reported and the increase with decreasing dimension is much less pronounced, having a power-law exponent of 0.34 (0.39 for calculated values) in contrast to 1.83 for the microwhisker experiments. The discrepancy to the microwhisker results may arise from the different measuring principles and the low accuracy in diameter measurement for this experiment [135]. But still, in contrast to metallic and non-metallic systems reported in this thesis, the observed yield strength is only a fraction of the calculated ideal strength (also for the corrected values). Deformation close to the theoretical limit is not observed because of the easy nucleation of dislocations. Further, the observed thinning and necking points towards a multiplication process that is active. This multiplication is preventing the crystal from starvation of mobile dislocations. As a consequence the crystal always contains defects that can interact with each other, switching off the size effect in these specimens.

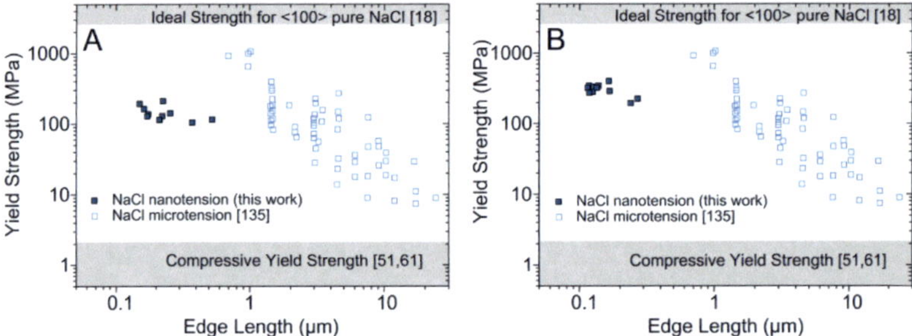


Figure 6.6: Masterplot of NaCl small scale mechanical measurements. The results of tensile testing of nanowires and are compared to values obtained in microtension [135]. (A) shows the measured data, whereas in (B) the cross section of the specimen was adjusted for the expected modulus.

# 7   Gold Nanowires

As a metallic system, Au nanowires were tested in tension. The growth of these specimens is described in chapter 1.6. Gold is a face-centered cubic (fcc) metal. It is noble and does not react with oxygen under ambient conditions. Since there is no native oxide formation, testing at small scales is expected to be readily interpretable. Bulk Au material is known to deform by the glide of full dislocations. It is very soft in its pure form with a yield strength only on the order of a few MPa.

## 7.1   Tensile Testing of Nanowires

Individual nanowires with diameters between 40 and 200 nm were tested in tension inside the vacuum chamber of the dual-beam microscope, according to the procedure described in detail in chapter 2.2. Strain was applied at quasi-constant rates of $10^{-4}$/s.

During the mechanical experiments, the nanowires exhibited elastic-plastic transitions at high stresses, typically exceeding 1 GPa. The detailed mechanical response and the corresponding deformation morphology showed significant variations. Characteristic stress-strain curves, obtained by force measurement in combination with DIC, are shown in Figure 7.1. After a period of linear-elastic loading, plastic deformation commenced by the formation of discrete slip steps. From this point, distinct differences were observed and the mechanical response can generally be divided into two classes. In the first class, plasticity occurred without signs of localization. Figure 7.1A shows the mechanical response for class I, where the initial yield point is followed by plastic deformation at the level of yield stress until the wires fracture between 2.5 and 5.5 percent total strain. During plastic deformation additional slip steps became visible in the SEM images. They appeared to be randomly distributed along the length of the wire, as can be seen in Figure 7.2C and D. In many cases only one active slip system could be identified (Figure 7.2C), however two active systems were observed as well (Figure 7.2E). Upon further straining, the number of slip steps increased and eventually the wires ruptured. Failure occurred abruptly, in most cases without signs of localization. Sometimes faint tensile necking could be observed (Figure 7.2D). The fracture surfaces indicate shear fracture as the predominant mode of failure.

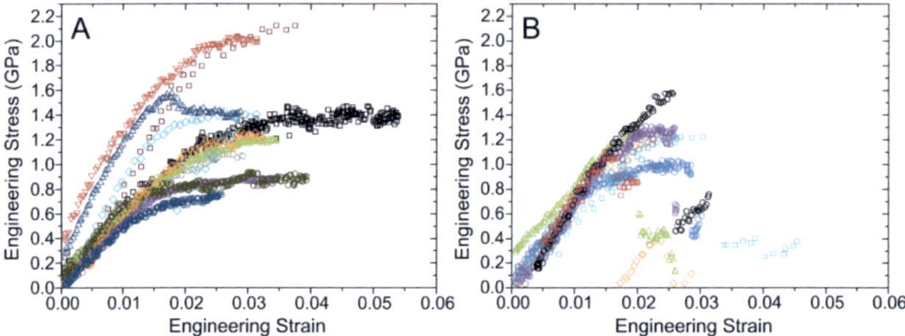

Figure 7.1: Characteristic stress-strain behavior of Au nanowires. (A) shows the of behavior class I where, after yielding, constant plastic flow on the level of yield stress was observed. In (B) the mechanical response of class II is visible. After yielding at stresses similar to those of the first class, the stress dropped to a significantly lower level where flow continued.

In class II, deformation was localized and extended thinned regions formed before failure. After yielding at high stresses, no plasticity or rather small amounts thereof were detected. Instead, the stress dropped at about 1 to 3 % of total strain and flow took place at significantly lower stress levels (Figure 7.1B). These stress-drops could be correlated to the formation of extended thinned regions that appeared within one or two frames (corresponding to 1-3 seconds) of the recorded movies. In most cases a slip step was visible shortly before the stress dropped. In the next image the region close to the slip step was deformed. After small amounts of further strain-

ing the wires fractured at the transition region of the deformed and undeformed portion of the wire. The deformed part then tilted to the side, forming a characteristic angle relative to the initial wire axis (Figure 7.2A and
B). Only one active slip system could be observed in these specimens. In
all cases the thinned region was several times larger than the width of the
wire. However, different lengths of the tilted region and different magni-
tudes of stress-drops were observed. Although no clear correlation between
both parameters was possible, the data indicates a gradual transition between the two aforementioned classes.

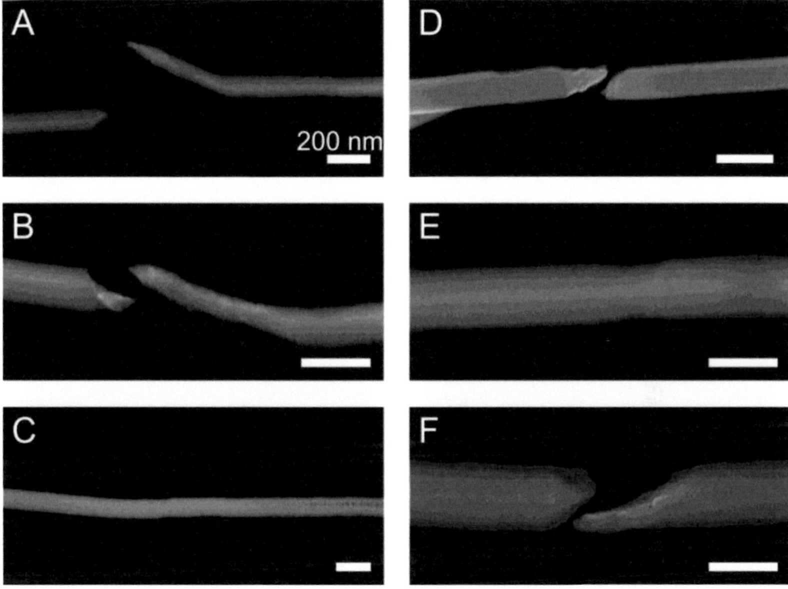



Figure 7.2: Deformation and fracture morphology of Au nanowires. (A) and (B)
represent the deformation mode of class II, where the stress drops and extended
thinned regions form. (C-D) show the deformation and fracture morphology of
class I. Small slip steps appear randomly distributed along the entire specimen.

In order to elucidate the underlying deformation mechanisms, EBSD analysis was performed on the fracture sites of eight wires, belonging to both cases. In all cases except one, the fractured region showed a strong change in crystal orientation at the fracture site. Figure 7.3 illustrates an EBSD investigation of a tested wire, belonging to the second class of deformation and to the purple tensile curve in Figure 7.1B. An SEM image of the deformed and tilted region after failure is displayed in Figure 7.3 and the transition to the deformed portion is marked by a dashed line. Figure 7.3B-D show detailed information obtained by EBSD. The inverse pole figure (Figure 7.3C) and the corresponding pole figure map (Figure 7.3B) indicate the [101] orientation along the wire axis (x-axis) with a small spread in orientation. For a better visualization a representative unit cell of each region is shown close to the transition. The deformed region appears as a twin in the EBSD data and has a spread of roughly 10 degrees between markings "2" and "4". Within these markings the orientation gradually changes, as can also be observed from the orientation distribution map in Figure 7.3D. As visible from Figure 7.3C, the spread in orientation is only strong around the z axis and rather weak perpendicular to it. Marking "3" represents a region, which has an orientation close to the orientation of the undeformed part of the wire. Therefore it either did not twin or has twinned twice to come back to its initial orientation. Figure 7.3E shows a TEM image of the region of failure, confirming the measurements obtained by EBSD. At the transition to the deformed zone the twinning plane is visible. Close to the fracture site, some more distinct twinning activity is observed. Strong bending contours are visible throughout the whole deformed region.

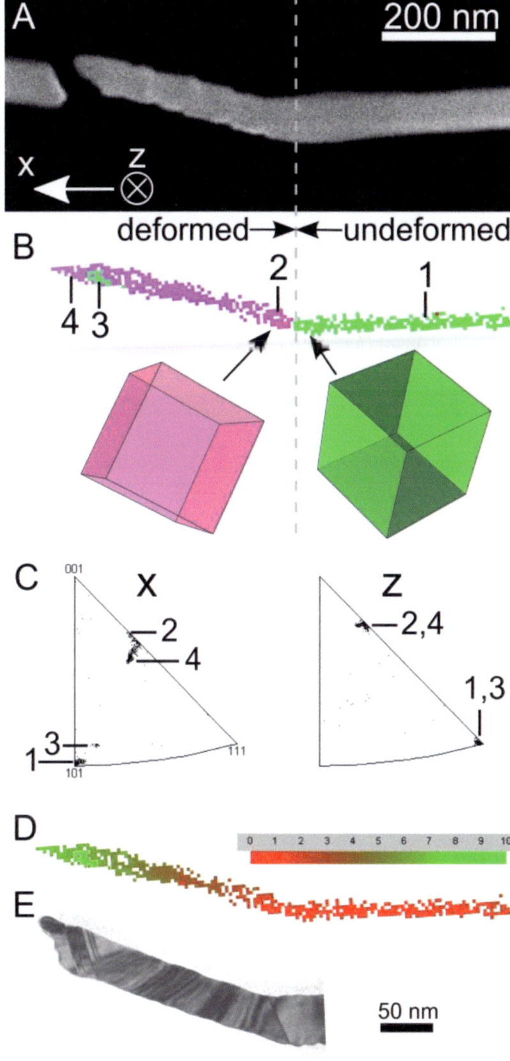



Figure 7.3: EBSD investigation of a deformed nanowire. A SEM image (A) shows the deformed and tilted part after fracture. The inverse pole figure map (B) and the inverse pole figure (C) show the [101] axial orientation of the nano-wire and identify the deformed zone to be a twin of the initial microstructure. Corresponding unit cells are used to visualize the orientations. Little spread in orientation is seen in the undeformed part, however for the twinned region, a gradual change in orientation of about 10 percent is obtained from the inverse pole figure (C) and the orientation distribution map (D). TEM investigations on the same specimen (E) confirm the observations made by EBSD. Strong bending contours are observed throughout the deformed region.

Figure 7.4, high-resolution TEM (HRTEM) images, taken by Gunther Richter at MPI-IS Stuttgart, corresponding to the specimen in Figure 7.2A, show twin lamellae, which formed during the deformation process. The fracture site is to the right of the images. The undeformed matrix of the nanowire is named M, whereas for labeling twin lamellae a T is used. Figure 7.4A shows twin embryos of roughly three layers spacing. In Figure 7.4B larger twins are seen, owing to the proximity of the strongly deformed area. All twins were identified as $\Sigma 3$ (111)[10-1] type symmetric tilt grain boundaries.

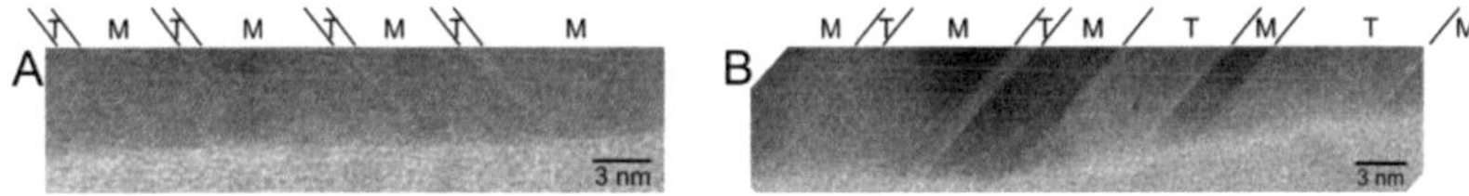


Figure 7.4: HRTEM images of the specimen from Figure 7.2A. (A) shows twin embryos of about three layers spacing. In (B) larger twins are observed in the vicinity of the strongly deformed region (HRTEM images courtesy of Gunther Richter, MPI-IS Stuttgart).

To confirm the deformation by twinning, the Eularian angles obtained by EBSD were used to plot the unit cells of the different regions, as already shown in Figure 7.3B. Figure 7.5A illustrates the superposition of the two unit cells (green = pristine, red = deformed). Additionally the potential twinning plane is inserted. If the entire structure is rotated in a way that the twinning plane is parallel to the plane of projection, the twin symmetry becomes evident.

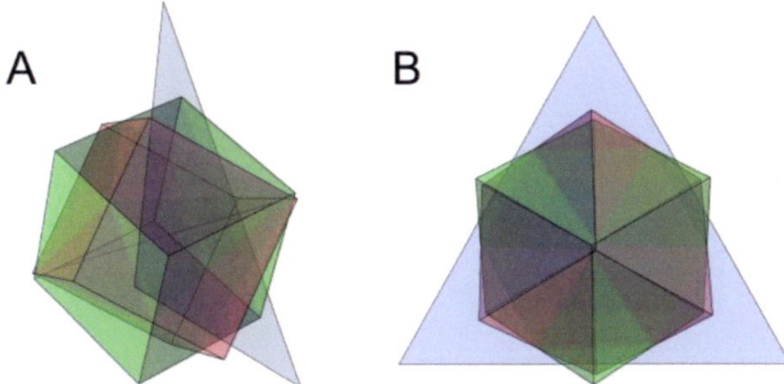


Figure 7.5: Superposition of unit cells (green = pristine, red = deformed) from Figure 7.3B and addition of the potential twinning plane (blue color). If the twinning plane is rotated to be parallel to the plane of projection, the twin symmetry is verified.

In one extraordinary case, the plastic flow could be sustained up to 12 percent of total strain. Figure 7.6 shows the stress-strain curve (Figure 7.6A) and the corresponding SEM images (Figure 7.6B). The wire followed the linear-elastic slope (I) until a small slip step could be seen and concurrently a deviation from linearity was observed in the stress-strain data. In the course of plasticity the stress dropped about 50% to 400-600 MPa. Simultaneously, the onset of plastic deformation was observed (II) manifesting itself by thinning of the wire and further extension of the thinned region to the right at the lower stress level (III). After fracture (IV), the deformed part of the wire tilted to the side, as has already been described before. TEM investigation of this nanowire (Figure 7.6C) revealed that the deformation occurred by multiple short twins on one glide system. It can be seen in the stress-strain curve that the wire was already under tension when the test started.

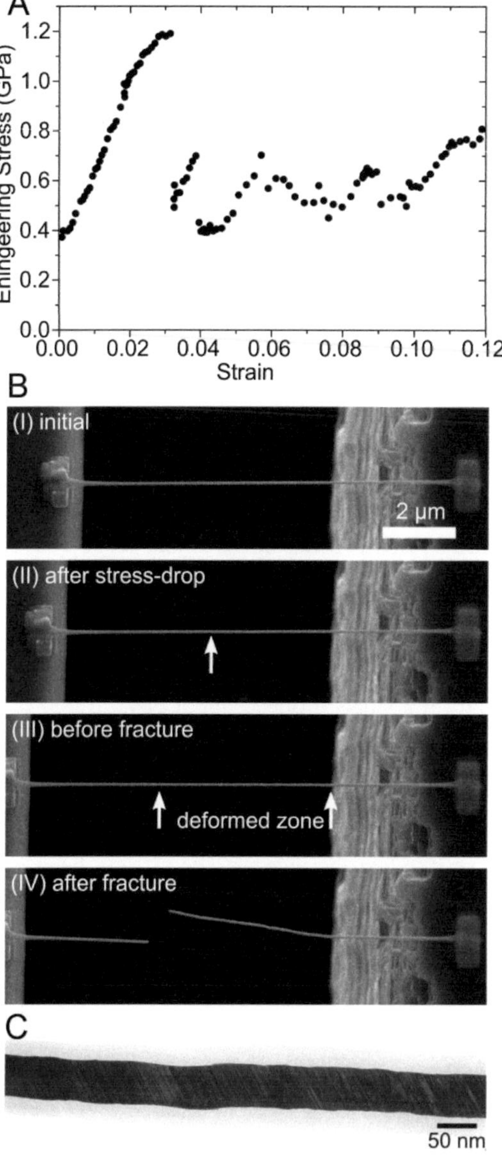



Figure 7.6: Pronounced plastic deformation of an Au nanowire. (A) shows the stress-strain curve. After yield the stress drops and flow occurs at a level that is approximately half of the yield strength. In (B) consecutive stages of plastic deformation are visible. The deformed region of the wire tilts to the side after fracture. TEM investigation shows many short parallel twins in the deformation zone (TEM image courtesy of Gunther Richter).

The growth of Au nanowires was not only confined to conventional Wulf shaped nanowires, like reported previously for Cu nanowhiskers [109] of the same synthesis method. During harvesting and manipulating single specimens, different cross-sectional geometries could be observed. For accurate calculations of the observed strengths, the fractured wires were bent in a way that their tensile axis was parallel to the electron beam. Subsequently the specimens were cut by the ion-beam outside the strongly deformed regions to obtain their initial cross sections. Figure 7.7 shows that the cross-sectional shape is not consistent throughout all specimens. Strong variations were found ranging from rhombic (Figure 7.7A), truncated rhombic (Figure 7.7F, Wulf shape) and rectangular (Figure 7.7C) to other modifications as trapezoid (Figure 7.7D) or bow-tie (Figure 7.7E). Most commonly for the nanowires were rhombic cross sections, were all side surfaces are {111} and truncated rhombic where two {111} surfaces are cut by a {100} surface. For the rectangular wires it is most likely that the long edges are low-energy {111} surfaces. On all cross sections varying amounts of contamination of the nanowire surfaces was visible. This layer is basically of carbonaceous origin. To some extent Pt is incorporated due to the method of fixation of nanowires using local EBID. Since this layer is distorting the stresses measured in the specimens, corrections can be made using a Voigt composite model (cf. chapter 2.2.3). Assuming no or negligible load bearing of the much weaker amorphous shell the thickness can be calculated. With this knowledge the stresses in the Au cores are obtained. Also, the cross sections of the nanowires can be accurately measured when bent up (Figure 7.7) leading to a similar result as using the Voigt model here. In principle the Voigt model could be used to check the accuracy of cross-sectional measurements in Au nanowires. In the end the measured and calculated values of stress in the nanowires provide a lower and a more accurate upper bound of strength.

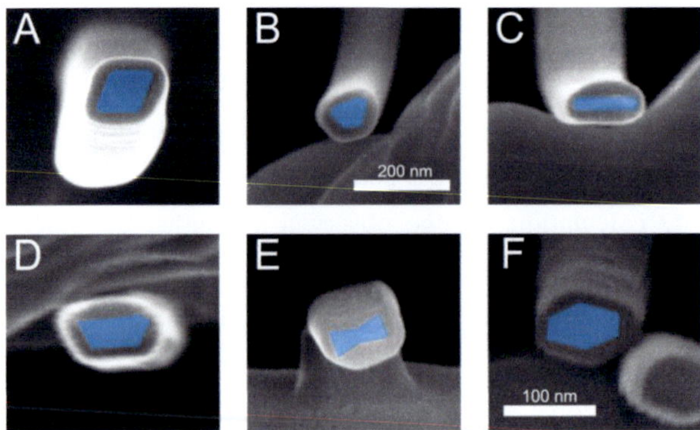

Figure 7.7: Cross sections of Au nanowires showing significant variations. The observed shapes ranged from rhombic (A), truncated rhombic (B) and rectangular (C) to other modifications as trapezoid (D) or bow-tie (E). The obtained geometries are highlighted in blue color. Surrounding the metal core highlighted in blue, strong contamination is visible.

## 7.2    Tensile Testing of Nanoribbons

The cross-sectional shapes of about a third of the specimens tested were rectangular with aspect ratios ranging from 2 to 8. Figure 7.8 shows characteristic SEM micrographs of these specimens which were named nanoribbons. When bent by or attached to a manipulator (Figure 7.8A and B) the ribbon-like structure and behavior can be seen. Figure 7.8C shows a rectangular geometry that was bent and imaged along the tensile axis. One edge length is about 10 times larger as the other edge length. Also a light layer (less brightness) is surrounding the nanoribbon (brighter area). This special structure was only observed when one edge length of the structure exceeded ~ 150 nm.

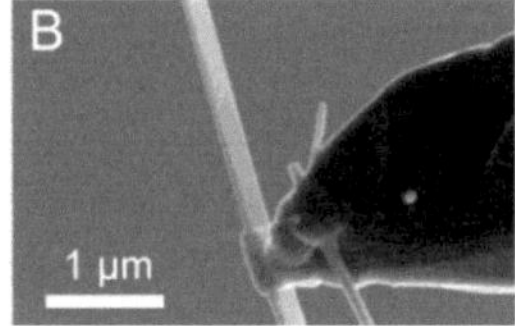
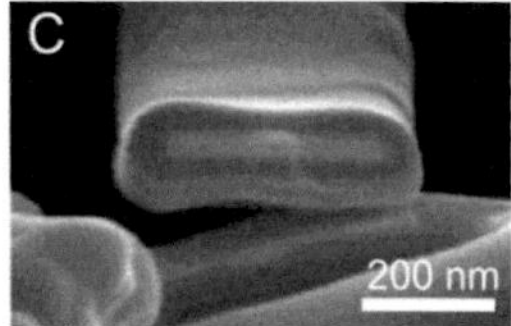


Figure 7.8: SEM micrographs revealing rectangular cross sections of nanoribbons. The ribbon like structure can be clearly seen when bent with a manipulator (A). In (B) the short and long edge are visibly distinguished. (C) shows the cut fracture surface of a ribbon, where the metal is surrounded by a contamination layer.

Tensile testing was performed on 8 nanoribbons. All showed high strength in the same regime as the nanowires. However, the stress-strain data of the experiments constantly exhibited the same characteristic shape, which can be seen in Figure 7.9A. After linear-elastic loading the ribbons yielded at varying stresses, which were not significantly different from the stresses observed in the nanowires. This was followed by a transition to ductile deformation at the yield level sometimes accompanied by different amounts of work hardening. The specimens sustained total strain between 3 and 10%. To confirm the elastic-plastic nature of deformation, as well as to allow for relaxation some ribbons were unloaded and then loaded again. During unloading no signs of plastic deformation were observed. For the stress-strain curve shown in Figure 7.9B, a ribbon was tested twice. After fracture one of the remaining parts was reattached and strained again. Plastic deformation of the second tests roughly starts at the stress level at which fracture had occurred before. Also the slope of strain hardening is consistent with the first test.

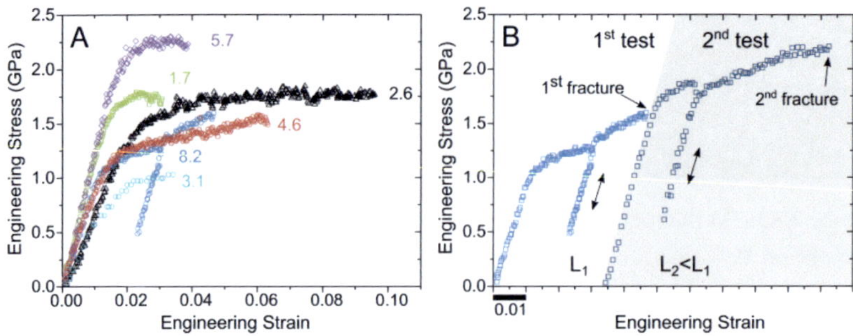

Figure 7.9: Tensile stress-strain curves of Au nanoribbons. Linear elastic loading is always followed by plastic flow with different amounts of plastic strain (A) and varying amounts of work hardening. (B) shows the consecutive test of a single nanoribbon. Plastic deformation of the second test approximately starts at the stress level where fracture had occurred before.

Concurrently imaging the ribbons during straining revealed the appearance of shear bands, which could be correlated to commence with yielding in the stress-strain data. Upon further straining the number of shear bands increased. They were found at various locations along the ribbons which seemed to be randomly distributed. In contrast to the nanowires, the slip traces were always found to be oriented in two different directions, often intersecting each other (Figure 7.10). Fracture took place by extending and shearing along one of the slip traces. In some cases extended shearing alternated between several sites (Figure 7.10A and B) until fracture occurred at one of them. In other cases tensile necking was observed (Figure 7.10C). EBSD was performed on fracture sites of ribbons which, like the nanowires, were determined to be twins of the undeformed regions. In contrast to the nanowires, the shear bands were always narrow never exhibiting extended thinned regions.

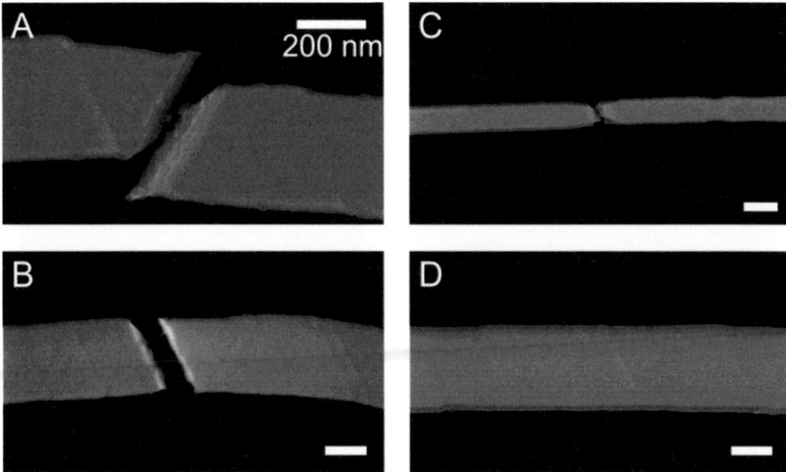


Figure 7.10: Deformation and fracture morphology of Au nanoribbons. Slip traces with different amounts of slip can be seen on the specimen surfaces. Slip always occurred on at least two conjugate slip systems, which caused the shear bands to intersect (B and D). In some cases tensile necking was observed (C). Fracture occurred by entirely shearing off at one of the slip steps.

Figure 7.11 shows the mechanical data of the Au nanowires and nanoribbons. It can be seen that there is large scatter in both the yield strength (Figure 7.11A) and the Young's modulus (Figure 7.11B).

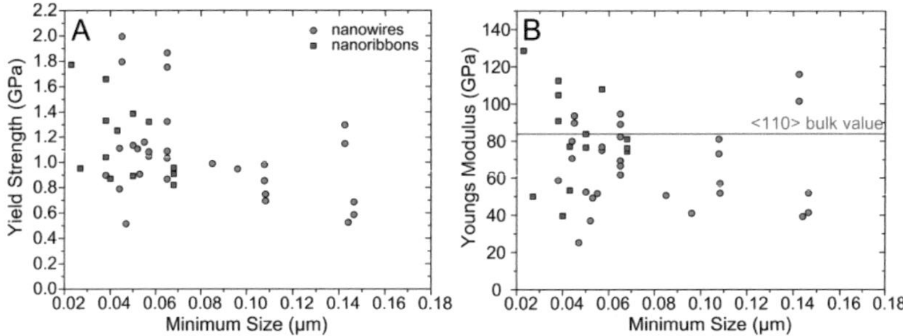

Figure 7.11: Mechanical data of Au specimens. Large scatter is observed for the measured yield strength (A) and the Young's modulus (B) for nanowires and nanoribbons.

## 7.3   Discussion

The tensile experiments on Au specimens show high strengths with large amounts of strain (3-12%). Further, a change in deformation mechanism from full dislocations to partial dislocation activity and deformation twinning is observed. The experiments show significant scatter in the stress-strain curves and the correlated deformation morphology, falling between the two extreme cases of class I with smooth flow at the level of yield stress and the formation of many slip steps distributed randomly along the wire length and class II with stress-drops after yielding, which could be correlated to the formation of extended thinned regions. The cross-sectional geometries were found to vary from rhombic or truncated rhombic to rectangular cross sections as already published for iron and silver whiskers by Brenner in the 1950s [177]. In contrast to the nanowires, nanoribbons having rectangular cross sections only showed the behavior of class I. The discussion is separated into three sub-chapters addressing deformation twinning, shape dependent behavior and the comparison to other small scale mechanical experiments on Au, as well as the application of weakest-link statistics on ductile nanomaterials.

## Twin Mechanism

For this first section complementary MD simulations were performed by Erik Bitzek at the University of Erlangen (Germany). Both, experiments and simulations, helped to reveal and to understand the two different mechanisms of twin deformation which can exist in Au nanowires. The results and many of the following arguments in this section were developed in close collaboration and parts were published in [178].

In the experiments twin deformation could be detected by means of EBSD measurements. The inverse pole figure map in Figure 7.3 shows the twin orientation of the deformed section to the initial wire matrix. In the coordinate system of the matrix, the wire axis changes from [101] to [-1-14]. If a new coordinate system along the tensile direction of the reoriented twin is applied the wire is reoriented along a <100> direction. TEM investigations corroborate the existence of both extended thinned regions that are fully twinned and multiple, two or more-layer, twin embryos in nanowires. From the SEM images one or two active slip-systems were observed during testing. No strong correlation was found between different cross-sectional shapes (rhombic or truncated-rhombic) and the modes of deformation, whereas significantly more wires with only one active slip system exhibited load drops. In order to gain deeper insights into the deformation mechanisms and to elucidate the observed differences in the experiments, complementary MD simulations were carried out by Erik Bitzek (University of Erlangen, Germany). The simulations were performed on specimens of experimentally observed shape [109]. As in other MD simulations conducted on fcc metals before [31], twinning was revealed as the governing deformation mechanism upon application of tensile stress along <110> directions. This process was observed regardless of the initial structure, the applied potential or the strain-rate. Furthermore, morphologies and trends in the stress-strain curves similar to the experiments could be observed. The change in orientation was identical to the experimental results and additionally the reorientation of surfaces could be observed. Nucleation of the first leading partial dislocation at the edges between two surface facets was always the first step of plasticity. The next partial then glided on an adjacent plane forming a two-layer twin embryo. At the intersections of this twin with the surface, further twinning dislocations were emitted, extending the twin boundary in a self-stimulating fashion. Propagation was always possible in both directions of the twin. Additionally, a second mechanism was identified, starting with the generation of twin embryos ahead of an extended twinned section. These then could grow and coalesce with each other and with the main twin, leading to accelerated twin growth. According to the way this mechanism manifested it was called leapfrogging.

The transition to deformation twinning, as observed in experiments and simulations, is not expected for Au at room temperature and experimental strain-rates. In the following paragraphs possible reasons will be addressed. Deformation in a defect free material always starts with the nucleation and passage of the first partial dislocation. The following process involves the nucleation of the second partial dislocation, be it a trailing, twinning or another leading partial dislocation on a different slip plane. One reason for subsequent twin deformation could be the difference in CRSS on leading and trailing partial dislocation [29, 31]. Karaman *et al.* [179] calculated the different Schmid factors for leading and trailing partial dislocations and argued that twinning can be crystallographically favored or not. Also subsequent stimulation of other leading partial dislocations, resulting in twin formation, may be promoted by intersections of partial dislocations with surface edges [24]. Additionally, energetics may play a role, since the nucleation of a twinning partial dislocation is favored if compared to the nucleation of a stacking fault at a different site, since no new fault area has to be created. The ratio of the stable stacking fault energy $\gamma_{sf}$ to the unstable stacking fault energy $\gamma_{usf}$ was used to explain full vs. partial dislocation activity [35] in nc materials. In the same context, twinning in fcc metals was modeled as the ratio between stable $\gamma_{tf}$ and unstable $\gamma_{utf}$ twin fault energy [37, 38, 180]. However, the different resolved shear stresses were not taken into account and only specific loading conditions were investigated [37, 38]. Following the calculations of the critical grain size $D_c$ for twinning in nc fcc metals [20], the Schmid factors for full ($b$) and partial ($b_p$) dislocations, as well as the corresponding Burgers vectors, $m$ and $m_p$, can be included to obtain the expression [178]:

$$D_c = \frac{2\alpha\mu\left(b\dfrac{m_p}{m} - b_p\right)b_p}{\gamma_{SF}} \tag{21}$$

with $\mu$ as the isotropic shear modulus. The factor $\alpha$ depends on the character of the dislocation and a scaling factor between the length of the dislocation source and the grain size. For the [101] oriented wire, the slip systems with the largest resolved shear stresses have a Schmid factor of $m = 0.408$. Their corresponding leading partial dislocation has a Schmid factor $m_1 = 0.471$, whereas the trailing partial dislocation the Schmid factor only is $m_2 = 0.236$. With $\mu = 22.55$ GPa, $b = 2.855$ °A, $b_p = 1.666$ °A, $\gamma_{sf} = 31$ mJ/m$^2$ [181] and $\alpha = 1$, the critical grain size according to equation 21 is on the order of $D_c = 40$ nm. Since the dislocation source size is usually much smaller than the size of the crystal, the true critical size for deformation twinning may therefore be exceeding this value many times [178].

For the reoriented <100> wire the slip system with the largest resolved shear stress has a Schmid factor of $m = 0.408$. However, the leading partial dislocation only has a Schmid factor of $m_p = 0.236$ leading to zero twin tendency, which is in agreement with the simulations [182] and experiments.

Although deformation twinning is seen in all specimens, two distinct modes of deformation are observed. The respective mode of deformation strongly depends on the nucleation sites available, the initial defect density and the surface roughness [178]. Also thermal activation may facilitate nucleation of partial dislocations [102]. The first class of deformation behavior is expressed by smooth yielding and the formation of nanotwins randomly distributed along the wire length. Instead of growing further, new nanotwins are created on multiple slip systems. This mode of deformation may also be assisted by the suppression of twin growth once a nanotwin is created. One mechanism seen in MD simulations is the interaction between growing twins on conjugate slip systems [182]. Another possible mechanism for this could be the superposition of bending stresses caused by the nanotwins, which would favor a different site for nucleation rather than continuing deformation at the initial location. In class II of deformation at least one twin is able to grow to larger sizes, which is accompanied by a change in mechanism. Rather than continuing propagation in a (conventional) layer-by-layer fashion, leapfrogged nanotwins on glide planes parallel to the twin plane are emitted ahead of the main twin, due to stimulation by the internal bending stress, which can be affected by several parameters such as applied load, length of the twinned section, crystal orientation and the lateral compliance of the testing system. Since the twinned section is reoriented and thus would naturally tilt away from the tensile axis (cf. Figure 7.2A and B) the internal bending stress is increasing with the size of the twin and there may be a critical twin size for the leapfrogging events to commence. Subsequent propagation causes these twins to coalesce with each other and the primary twin boundary, leading to self-sustained and accelerated growth of extended twins. A morphological feature of this mechanism is the large twin, which tilts away from the wire axis after fracture (Figure 7.2A and B) and can be easily confirmed by EBSD (Figure 7.3). Also the strain-rate at which the experiment is conducted could play a role to determine, whether a long twin can form or not. Seo *et al.* [33] used a piezo based system, which employs the slip-stick mechanism, mentioned in chapter 2.1.1, for the application of displacement. Their wire underwent twinning throughout the whole specimen. A possible explanation is the strong variation in strain rate during the tensile test, caused by the coarse mode slip-stick mechanism. The lateral compliance of the testing setup was higher, so the bending stresses due to the reorientation and subsequent tilting of the twinned section could be better accommodated. Also in this work variations in strain-rate may have been existent, which could cause differences in the mode of deformation. For the transducer employed here, forward surges were detected (Figure A.1.1) Due to the inability of the device to control these fast events of elastic energy release, an increased strain-rate was applied upon the observed stress-drops until control was regained. Detailed information can be obtained from Appendix A.1.3. Hence the creation of long twins could also be facilitated by a higher strain-rate.

The experiments show significant scatter in the stress-strain curves and deformation morphology, all lying in between the two extreme cases, which may be a consequence of the statistical nature of the deformation processes and point towards competing mechanisms that both are active to a certain extent. In some cases one twin can grow to a larger size, if it does not encounter an obstacle during this process. Once grown large enough the mode of deformation changes and a long twin is formed. The randomness of this event can be shown by subsequently testing a single nanowire. Figure 7.12 shows that both modes can exist in one specimen.

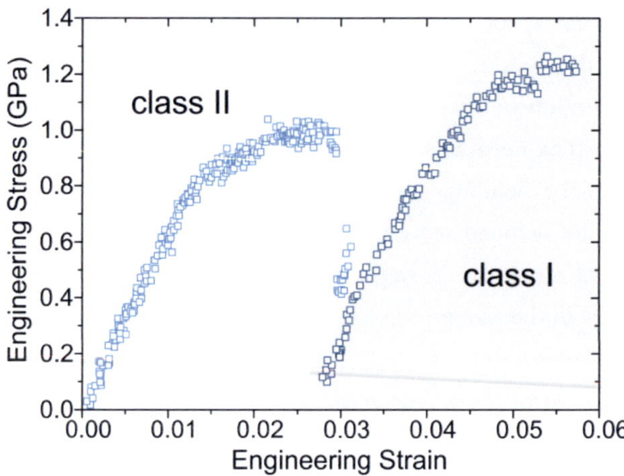


Figure 7.12: Consecutive testing of a single Au nanowire shows the randomness of deformation. Both classes of deformation can exist in one specimen.

Further evidence for this randomness is provided by the observations of Figure 7.6, which shows one extreme case, where the wire was able to sustain a total strain of 12%. Upon yielding a stress drop occurred, which could be correlated to the formation of a slip step in the SEM image. During further straining at the lower stress level, a thinned region extended throughout the wire. As more than one third of the wire was deformed, fracture took place at one of the interfaces and the deformed region tilted to an angle, which was smaller than observed before. TEM investigations revealed highly coordinated growth of a large number of nanotwins on atomically similar twinning systems having small spacings of the undeformed matrix between them. In this case, the wire was able to sustain larger amounts of strain since it did not precociously encounter an obstacle, like a twin on a conjugated slip plane. It is also possible that the superimposed bending stress could further be balanced by the undeformed sections between the small twins. This type of deformation only can happen if just one system for the nucleation and propagation of partial dislocations is active. A partial dislocation from a conjugated slip system would intersect with the twin and thus hinder plastic deformation.

## Shape Dependence

A significant amount of specimens among the Au nanowires was found to have a rectangular cross section and from there on they were dubbed nanoribbons. In contrast to the nanowires this type of nanostructure solely shows the "plateau-like" mechanical behavior of class I. Only small slip traces are visible on multiple slip systems (Figure 7.2). For these rectangular specimens there is a short and a long edge of the cross section (Figure 7.8C). By placing the nanoribbons on the substrate during manipulation the long edge always happens to lie more or less flat on the surface. Thus the view on the fractured specimens in Figure 7.10 delivers direct information on this long face of the nanoribbons. In all cases it can be seen that at least two different slip systems are active. The orientation of the ribbons was determined to be similar to that of the nanowires by means of EBSD. Measurements of the angles between the tensile axis and the slip traces on the surface scatter around $60°$. The scatter can be produced by the accuracy of the measurement or most likely by variations in the alignment, since in the 2D view on the top face of the wire tilt in the specimen will change the measured angle. With the details about the growth of these specimens [109] the side surface orientations have to be either $\{111\}$ or $\{100\}$. Since the orientation of the specimens is known, as is the glide plane in fcc metals, a slip trace on a $\{100\}$ surface would have to appear perpendicular to the $<110>$ tensile direction on the nanoribbon surface, whereas the normal to a $\{111\}$ slip plane exhibits an angle of $\sim35.26°$ to the $<110>$ direction revealing the long side faces to be $\{111\}$.

The deformation mechanism is still DT in nanoribbons, since EBSD confirmed the twin symmetry of the deformed sections. A transition to ODP due to the larger size of the ribbons can therefore be disqualified. The observed difference could be caused by the different lengths of the paths that the partial dislocations have to travel in nanoribbons. With the longer distance there is a higher probability that the leading partial is caught up by a trailing partial dislocation or runs into another leading partial dislocation or stacking fault on a conjugated slip plane. In both cases twinning at this site is terminated. The high density of intersecting slip traces points towards a strong interaction between nanotwins in the ribbons. Once the growth of a twin is stopped by a partial dislocation on a conjugate slip system, a higher stress is needed to activate the next system, confirmed by the significant amount of work hardening in Figure 7.9B. The curve fit between the first and the second test also reveals the statistical nature of this process. In this specimen alternating reactivation of abandoned sites could also be observed at increasing stresses. A second possible explanation for the existence of only class I behavior involves the geometrical differences between wires and ribbons, in which a difference in the moment of inertia between the y and z directions exists (Figure 7.8). According to the slip traces, the ribbons would have to tilt over the short edge to accommodate twins of significant size, but instead the higher bending stiffness in this direction superimposes stresses that stop twinning and make the nucleation of leading partial dislocations more favorable on other sites. Although no correlation could be found between yield strength and measured aspect ratio of the side faces, there may be a critical twin size for each aspect ratio to stop propagation and start nucleation at a different location.

Fracture in both geometries is expected to take place at obstacles, e.g. the intersections of twins on conjugate slip systems, which is clearly seen in simulations [178, 182]. In all cases shear fracture was the predominant mode of failure, sometimes showing various degrees of tensile necking. These necks could possibly be formed by continuous nucleation of stacking faults at the already deformed region, subsequently thinning the region further, which eventually results in failure.

## Size Effect and Weakest-Link Behavior

Significant scatter is observed in the yield strength of nanoscale gold. Also the measured Young's modulus scatters well around the bulk value calculated for Au along <110> directions, indicating corruption in the mechanical data, since there is no good physical reason for the elastic properties to differ from that of bulk material. These falsifications are most likely due to errors in cross-sectional measurements, deriving from the contamination layer caused by e-beam imaging, which is further described in chapter 2.2.3. For accurate cross-sectional measurements, the wires were cut with the ion-beam after fracture and tilted towards the electron-beam for the determination of the cross-sectional shape and area. Layers of various amounts can be seen in Figure 7.7. The TEM image in Figure 2.12C reveals the core-sheath structure of the (heavy) Au nanowire, which is surrounded by (light, because electron transparent) contaminant. This can lead to erroneous area measurements, which in turn affect the measurement of stress in the specimens. The impact of contamination is increasing with increasing layer thickness and decreasing wire diameter, as shown in Figure 2.13B. As described in chapter 2.2.3, one can calculate the size of the contamination layer by applying a simple Voigt composite model, which is valid if the strain in both layers is assumed to be equal. The corrected stress in the Au nanowire is subsequently determined by using the calculated cross section. Comparing corrected strength values to the initial measurements is accompanied by a decrease in scatter, as shown in Figure 7.13.

In the context of mechanical size effects, which have been observed in micropillar experiments, the data is compared to other mechanical experiments on pure single-crystalline Au, such as micropillar compression [112, 114], microtension of FIB-milled specimens [183] and tensile experiments of microwhiskers [184]. In Figure 7.13 the results of the aforementioned experiments are combined into a masterplot with the measured values (Figure 7.13A) and corrected for the contamination layer (Figure 7.13B).

As mentioned in chapter 1.5, the FIB prepared specimens show a strong size effect with decreasing sample size with a power law exponent around -0.6. The exponent for the Au nanowires is found to be -0.26 for the corrected values (-0.3 for the initial measurements) and -0.19 for the nanoribbons (-0.39), which is significantly lower, although these values still suggest a size effect.

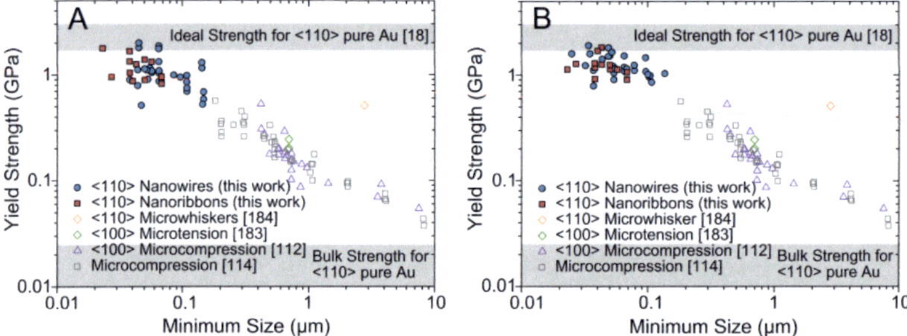


Figure 7.13: Masterplot of Au small scale mechanical measurements. The results of tensile testing of nanowires and nanoribbons are compared to values obtained in microcompression and microtension. In contrast to a power law exponent of about -0.6, values of -0.2 to -0.3 are measured in this work suggesting a less pronounced size effect. In (A) the measured values are plotted for wires and ribbons, in (B) these values are corrected for a contamination layer (cf. chapter 2.2.3).

Explanations for the different observations in nanowires and other small scale experiments can be drawn from different sides. It is possible that for the nanowires there is in fact no size effect. With this strong amount of scatter, which could possibly be caused by small irregularities as the absence of single atoms on the surface, the slopes measured from the linear fit of the yield strength data are very uncertain and have to be treated with care. The high purity of the specimens causes deformation at or close to the lower level of the theoretically calculated yield strength. As a consequence all specimens with this low defect density should show this behavior regardless of size. Second, the values corrected for the contamination layer in

Figure 7.13B and the findings of a transition in deformation mechanism elicit another possibility for these observations. If the slope in Figure 7.13B is extended, the exponent would rather fit to the data point obtained by Brenner who tensile tested similar specimens in the micrometer regime [184]. Although he tested 15 specimens, he unfortunately only presented one tensile curve and therefore this finding cannot be based on strong statistics. Furthermore, he did not report on the underlying deformation mechanism. If one assumes deformation twinning for all Au nanowires it is possible that the different behavior arises from the different mechanisms of deformation in nanowires and micropillars (DT vs. ODP). A third hypothesis is, that the difference in mechanical behavior is caused by the difference in crystal orientation. As stated above the tendency of an fcc metal to twin is strongly dependent on the crystal orientation in the loading direction determining the resulting Schmid factors for leading and trailing partial dislocation. Micropillar compression was always performed on <100> columns or columns oriented for single-slip, which are not prone to DT. A reversal in loading direction is also expected to activate different mechanisms in certain orientations. In contrast to this, a reversal in loading direction in <100> oriented micropillars was not found to affect the deformation behavior of FIB-milled specimens [183].

The two proposed mechanisms for the mechanical size effect, dislocation starvation [112] or source truncation [118], have exclusively been investigated for the case of ODP based deformation in crystals containing initial defects. The situation for the single crystalline nanowires is different. Here an extreme case of "starvation", namely no dislocation, exists from the beginning. This and the crystal orientation in combination with the loading direction cause that the strength and the mode of deformation in defect-free nanowires primarily hinge on flaws and surface roughness, which dictate the nucleation stress of the first partial dislocation and possibly the pattern and distribution of deformation twins along the nanowire length. For this type of argumentation a Weibull approach should reveal a dependence of yield strength on the geometrical dimensions of the specimens. Figure 7.14 shows the Weibull plots of yield strength vs. specimen volume, surface area and length. Although there is significant scatter, a faint dependence of yield strength on wire dimensions is visible, fortifying a higher probability of facile nucleation sites with increasing wire length. Further, an increase in scatter can be observed with decreasing sample dimension, confirming the statistical nature of events. A Weibull analysis was performed, according to the equation derived in appendix A.4, for nanowires and nanoribbons with both measured and calculated values of yield strength. The Weibull moduli for surface area and volume are shown in table A.1. For the measured values the moduli are significantly lower for wires, which could be due to the stochastic occurrence of two distinct twinning-mediated deformation mechanisms. However, if the calculated values for yield strength are used the Weibull modulus is increasing and the values for nanowires and nanoribbons converge. In all cases the Weibull modulus is falling between 3.4 and 6.6, which is very low and rather describing the behavior of a brittle material. For a soft metal like Au, showing deterministic behavior as a bulk material, a modulus well above 20 is expected. This is again manifesting the inherently stochastic behavior at these length scales, even for a material that shows a very narrow distribution in the macro regime, substantiating the tremendous effect of single flaws in a material with high crystalline perfection.

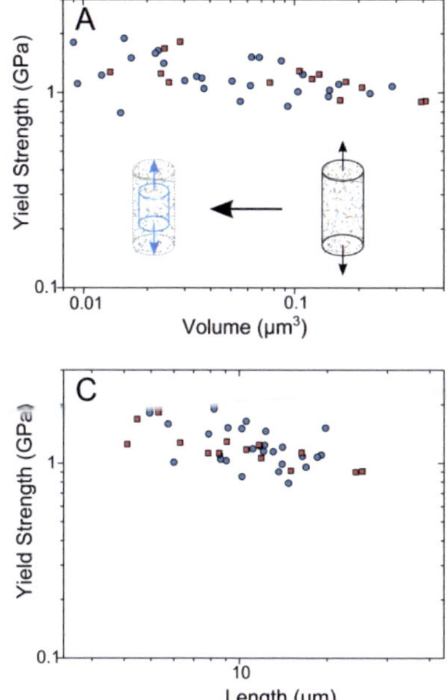

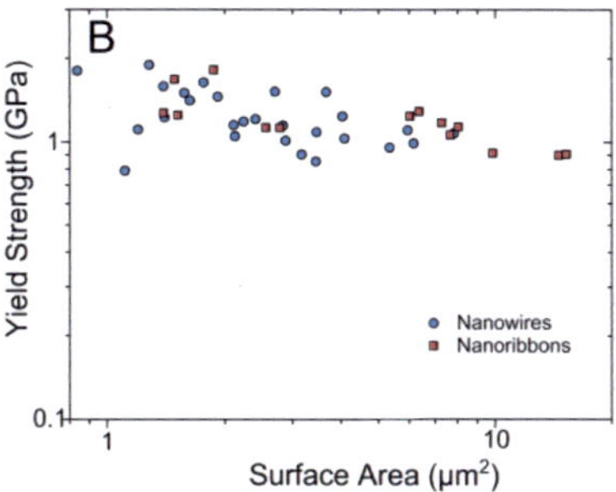


Figure 7.14: Weibull plots for Au nanowires and nanoribbons show the yield strength (corrected for a contamination layer) of the specimens in dependence of the geometrical dimensions of volume (A), surface area (B) and length (C).

# 8  Summary & Findings

The mechanical response of nanostructures as building blocks for micro- and nanoscale devices and nanostructured materials is key to their structural integrity and therefore of tremendous importance for their reliability. Recent studies through compression on a variety of materials have shown increasing strength with decreasing sample dimensions down to 100-200 nm. This "mechanical size effect" has gained substantial interest beyond the community of small scale mechanics. Even higher strength under tension at the same length scale, in experiments that date back to the 1950s, have inspired the work of this thesis – the investigation of the mechanical response of materials in smaller size regimes subjected to tensile stresses approaching the theoretical strength of the material.

In this work an experimental setup was developed to test quasi 1D nanostructures under monotonic uniaxial load. Nanowires of different crystal structures and atomic bonds were tested in tension. Concurrently the deformation of the specimens was observed and correlated to the mechanical response. Further post-mortem investigations were carried out to elucidate the underlying deformation mechanisms.

Two distinct setups for tensile testing were designed accompanied by a testing route that can be employed inside the vacuum chamber of electron microscopes. The quality and the throughput of measurements strongly hinges on the degrees of freedom provided by the particular testing suite. Further it is shown that the transfer into vacuum environments deteriorates the noise characteristics and therefore the smallest measureable quantity. Differences in actuation and measurement principles further influence the noisefloor. Enhancement can be provided by rigid configuration and the employment of suitable control algorithms. The compliance of the device itself and the material used for gripping as well as the contamination of surfaces that always accompanied irradiation require thorough interpretation of the data. Both testing platforms allow for the investigation of a variety of nanoscale materials and may someday be replaced by the ideal device for mechanical testing – an infinite stiff testing apparatus.

The variety of materials tested showed distinct behavior at the nanoscale. Table 8.1 compares the behavior revealed during tensile testing of nanowires with the conventional behavior known from bulk material.

|  | bulk | nano |
|---|---|---|
| **Mo** | ordinary dislocation plasticity | ordinary dislocation plasticity <br><br> • dependence on the character of single defects <br> • dependence on the frequency of single defects <br> • increase in strength and scatter |
| **VO$_2$** | phase transitions (MIT, IIT) | phase transition (IIT) can be quantified <br><br> • constant plateau stress <br> • increase in strength and scatter |
| **Si** | brittle (weakest-link behavior) | brittle OR ductile <br><br> • brittle in <111> and <100> if low doped <br> • increased strength and weakest-link behavior <br> • ductile if highly doped |

| **NaCl** | brittle<br>plasticity under compression<br>emission of dislocations from crack-tips | shear vs. cleavage<br>size-effect switched off<br>dislocation nucleation and multiplication and subsequent cleavage |
| --- | --- | --- |
| **Au** | ordinary dislocation plasticity<br><br>• deterministic behavior | partial dislocations, deformation twins<br><br>• orientation dependence<br>• shape dependence<br>• stochastic behavior |

Table 8.1: Comparison of the findings for tensile testing at the nanoscale with the conventional behavior known for the respective bulk material.

The Molybdenum fibers can be regarded as a proof of concept. Data of micropillar compression on the same material is available as a reference. For the yet relatively large (d=300-500nm) fibers the deformation pathway is still governed by ODP, as in bulk Molybdenum. In contrast to bulk the as-grown molybdenum fibers reach the theoretical limit of strength, as observed in pillar compression before. A significant change is the difference in scatter in the yield strength of as-grown fibers between tension and compression. The difference in gauge lengths causes a higher probability of finding a critical defect in the significantly longer tensile specimens. Highly prestrained fibers behave similar in compression and tension. Owing to the high dislocation density each specimen contains defects and deterministic behavior with little scatter is observed. The transition from linear-elastic behavior at the ideal strength, which is expected for a defect-free material, to smooth continuous plasticity, as seen in bulk material, is found in specimens with intermediate prestrain. They return both behaviors as well as intermittent plasticity due to the inhomogeneous structure of dislocations inside the material. Strain softening and the propagation of Lüders bands is observed in a limited number of tests, usually in E4 specimens, where the stored elastic energy is low enough for the PID control to work properly. These findings show that although the deformation mechanism is still valid the mechanical behavior of Mo fibers is governed by the existence of initial defects and their character.

If the dimension of the specimens is then decreased, as it is the case for the remaining materials tested, unexpected effects that differ from the behavior of bulk materials emerge. Bulk Vanadium dioxide shows phase transformations with temperature and mechanical load. In the single crystalline $VO_2$ nanowires the reversible IIT phase transformation can be shown and be quantified. The constant plateau stress is used to proof the accuracy of the tensile setup. Increasing strength and scatter breed a weakest-link argument for interpretation. Using electrical measurements piezoresistivity in the M2 phase is discovered and quantified. If measureable at all, these findings would have been strongly distorted in bulk polycrystalline specimens, foreclosing interpretation.

Silicon nanowires show similar and different behavior if compared to their bulk counterpart. Low doped <111> and <100> oriented nanowires exhibit a higher level of strength than bulk Si, but still show a strong variability and linear-elastic behavior. For these brittle materials the weakest-link approach is suitable to understand the observed behavior. Large differences are observed for different methods of growth, mostly owing to different frequencies and character of impurities or flaws. As a consequence only clean wires will show consistently high strength at the theoretical limit. If high doped a significant change from the behavior of bulk Si in the mechanical response is observed. Upon straining the stress-strain curve deviates from the linear-elastic slope suggesting ductile deformation in highly doped <100> oriented nanowires up to 22% of total strain, further supported by the remaining deformation after fracture. It is possible that doping can induce porosity in the specimens causing the observed transition. However the origin is still not fully understood and is calling for further investigation. This transition is harboring the chance to use Si in fields where its application has been excluded before.

Engineering stress-strain curves of Sodium Chloride nanowires show almost elastic-ideal-plastic behavior. The constantly ductile deformation is in strong contrast to our notice from the handling of bulk material (table salt). In this ionic structure, having glide and cleavage planes, the high crystal purity and clean surfaces cause slip to be favored to cleavage. Although pure single crystalline, the nucleation of dislocations requires relatively low stresses and in contrast to all other materials tested in this work, the behavior is deterministic. Once plasticity commenced a competition between shear and brittle fracture is started. The interaction and multiplication of dislocations provides a mechanism to avoid starvation and therefore continue plastic deformation but also causes hardening that can terminate plastic deformation. A defect, such as a surface flaw or a shear offset, at a different location of the specimen then can initiate a crack and cause brittle fracture. It is shown that the conventional continuum mechanics model for the determination of the critical flaw size is possibly not valid in nanoscale materials. Also the stress intensity factors may be different.

In Gold nanowires full dislocation plasticity is replaced by the nucleation of partial dislocations and subsequent deformation twinning. Whether DT occurs or not strongly depends on the orientation and the loading direction of the crystal and will only appear if the leading partial dislocation exhibits a higher Schmid factor than the trailing partial dislocation. Once a critical twin size is met the mode of propagation is accelerated by leapfrogging and paralleled twin growth. As a result the stress-strain curves show constant flow at the level of yield or stress-drops and flow at lower regimes. Nanoribbons of the same orientation in turn only exhibit the former mode of deformation, which shows that proper cross-sectional shape is able to suppress accelerated twin growth. The mode of deformation in Au nanowires hinges on initial defects and the probability of a propagating deformation twin to encounter an obstacle during growth or not. The yield strength of these nanoscale Au specimens is highly stochastic, requiring Weibull theory to be employed for interpretation in a material that is inherently deterministic in bulk.

A key issue in all tested materials, commonly observed at these small scales, is the measurement of the elastic properties. In all cases the modulus encountered strong variations, which could not be explained by a single parameter. Distinct origins for different specimens further complicate the elucidation. Overall it is assessed that the variation in Young's moduli of one single type of specimens can be affected by (i) misalignment, (ii) surface layers, (iii) the compliance of grips and (iv) erroneous determination of cross sections.

The lower bound of the calculated theoretical strength is reached for all investigated materials, except NaCl. The measured yield/fracture stresses are scattered and only for the Mo fibers a clear size effect is visible, which is not coupled to the lateral dimensions of the specimens, but to the length and the probability of finding a critical defect. In all cases the behavior becomes chaotic and unpredictable, since the yield or fracture strengths are affected by even small changes, e.g. the absence of a few atoms on the surface of the crystal can already alter the mechanical response. In summary the transition from deterministic to stochastic events requires to use weakest-link type approaches also for ductile specimens in this size regime. Distinct mechanical behavior is observed in all classes of nanowires sometimes strongly deviating from the behavior of their bulk counterparts. Often, a competition between modes of deformation, e.g. partial dislocations vs. deformation twinning in Au nanowires, is observed. At this size scale the borders of strict classes of mechanical behavior fade, possibly burying the chance for engineers to use e.g. semiconductors where their linear-elastic, brittle characteristics formerly excluded their application. However it has to be elucidated, whether the remaining physical properties can still be regarded as valid.

For engineers there are two ways of obtaining a defect-free and therefore strong crystal, (i) bottom-up growth or (ii) scaling down of matter. The stacking of individual layers of atoms (i) returns very pure structures with few defects. Also making a large crystal sufficiently small (ii) purifies a crystal and hence strongly lowers the defect density. As seen from the results of this work, both approaches still harbor a certain possibility of finding a critical defect, which governs the mechanical response of the respective specimen. Although the bottomline of strength and toughness is way above the bulk properties the materials are still exposed to strong variations that can already be caused by small changes as a few missing atoms.

The findings of this work show that it indeed is possible to tailor the mechanical response of a crystal with shape, crystal orientation, surface condition, initial defect density and possibly doping. With using the right parameters also inherently brittle materials can be tuned to fail graceful. The trade-off for this opportunity is the strong variation in mechanical response. Only with the ability of engineers of adding and removing single atoms to a structure will it be possible to obtain deterministic behavior at the level of the theoretical strength of a material.

With this thesis the groundwork for the investigation of quasi 1D nanostructures is provided. A departure from deterministic, bulk-like behavior is seen for almost all materials, accompanied by the competition of deformation mechanisms. The construction of commonly applied maps showing the mode of deformation vs. size are no longer possible. In this size regime, the size scale has to be replaced by a more appropriate parameter as orientation, shape or defect structure or some parameter that may not be resolved yet. New technologies as Helium ion microscopes (HIM) enable the removal of very small quantities of atoms allowing the determination of materials parameters as stress intensity factors at the nanoscale.

# Bibliography


[1]     R.P. Feynman, *There's Plenty of Room at the Bottom.* Sci. Eng., 1960. **23**: p. 22-36.

[2]     M.D. Uchic, D.M. Dimiduk, J.N. Florando and W.D. Nix, *Sample dimensions influence strength and crystal plasticity.* Science, 2004. **305**(5686): p. 986-989.

[3]     J.M. Tarascon and M. Armand, *Issues and challenges facing rechargeable lithium batteries.* Nature, 2001. **414**(6861): p. 359-367.

[4]     C.M. Lieber and Z.L. Wang, *Functional nanowires.* Mrs Bulletin, 2007. **32**(2): p. 99-108.

[5]     Y. Cui, Z.H. Zhong, D.L. Wang, W.U. Wang and C.M. Lieber, *High performance silicon nanowire field effect transistors.* Nano Letters, 2003. **3**(2): p. 149-152.

[6]     F. Qian, S. Gradecak, Y. Li, C.Y. Wen and C.M. Lieber, *Core/multishell nanowire heterostructures as multicolor, high-efficiency light-emitting diodes.* Nano Letters, 2005. **5**(11): p. 2287-2291.

[7]     Q. Wan, Q.H. Li, Y.J. Chen, T.H. Wang, X.L. He, J.P. Li and C.L. Lin, *Fabrication and ethanol sensing characteristics of ZnO nanowire gas sensors.* Applied Physics Letters, 2004. **84**(18): p. 3654-3656.

[8]     M. Law, L.E. Greene, J.C. Johnson, R. Saykally and P.D. Yang, *Nanowire dye-sensitized solar cells.* Nature Materials, 2005. **4**(6): p. 455-459.

[9]     C.K. Chan, H.L. Peng, G. Liu, K. McIlwrath, X.F. Zhang, R.A. Huggins and Y. Cui, *High-performance lithium battery anodes using silicon nanowires.* Nature Nanotechnology, 2008. **3**(1): p. 31-35.

[10]    M. Armand and J.M. Tarascon, *Building better batteries.* Nature, 2008. **451**(7179): p. 652-657.

[11]    M.D. Uchic, P.A. Shade and D.M. Dimiduk, *Plasticity of Micrometer-Scale Single Crystals in Compression.* Annual Review of Materials Research, 2009. **39**: p. 361-386.

[12]    F. Ostlund, K. Rzepiejewska-Malyska, K. Leifer, L.M. Hale, Y.Y. Tang, R. Ballarini, W.W. Gerberich and J. Michler, *Brittle-to-Ductile Transition in Uniaxial Compression of Silicon Pillars at Room Temperature.* Advanced Functional Materials, 2009. **19**(15): p. 2439-2444.

[13]    S.T. Boles, E.A. Fitzgerald, C.V. Thompson, C.K.F. Ho and K.L. Pey, *Catalyst proximity effects on the growth rate of Si nanowires.* Journal of Applied Physics, 2009. **106**(4).

[14]    M. Schamel, C. Schopf, D. Linsler, S.T. Haag, L. Hofacker, C. Kappel, H.P. Strunk and G. Richter, *The filamentary growth of metals.* International Journal of Materials Research, 2011. **102**(7): p. 828-836.

[15]    G.E. Dieter, *Mechanical metallurgy.* 3rd ed. ed. 1986, New York: McGraw-Hill.

[16]    G. Gottstein, *Physical foundations of materials science.* 2004, Berlin ; London: Springer. xiv, 502 p.

[17]    J.P. Hirth and J. Lothe, *Theory of dislocations.* 2nd ed. ed. 1982, New York ; Chichester: Wiley. xii,857p.

[18]    S. Ogata, J. Li, N. Hirosaki, Y. Shibutani and S. Yip, *Ideal shear strain of metals and ceramics.* Physical Review B, 2004. **70**(10).

[19]    M.A. Meyers, O. Vohringer and V.A. Lubarda, *The onset of twinning in metals: A constitutive description.* Acta Materialia, 2001. **49**(19): p. 4025-4039.

[20]    M.W. Chen, E. Ma, K.J. Hemker, H.W. Sheng, Y.M. Wang and X.M. Cheng, *Deformation twinning in nanocrystalline aluminum.* Science, 2003. **300**(5623): p. 1275-1277.

[21]    X.Z. Liao, Y.H. Zhao, S.G. Srinivasan, Y.T. Zhu, R.Z. Valiev and D.V. Gunderov, *Deformation twinning in nanocrystalline copper at room temperature and low strain rate.* Applied Physics Letters, 2004. **84**(4): p. 592-594.

[22]    S.H. Oh, M. Legros, D. Kiener, P. Gruber and G. Dehm, *In situ TEM straining of single crystal Au films on polyimide: Change of deformation mechanisms at the nanoscale.* Acta Materialia, 2007. **55**(16): p. 5558-5571.

[23]    H. Rosner, J. Markmann and J. Weissmuller, *Deformation twinning in nanocrystalline Pd.* Philosophical Magazine Letters, 2004. **84**(5): p. 321-334.

[24]    Q. Yu, Z.W. Shan, J. Li, X.X. Huang, L. Xiao, J. Sun and E. Ma, *Strong crystal size effect on deformation twinning.* Nature, 2010. **463**(7279): p. 335-338.

[25]    R.J. Asaro and S. Suresh, *Mechanistic models for the activation volume and rate sensitivity in metals with nanocrystalline grains and nano-scale twins.* Acta Materialia, 2005. **53**(12): p. 3369-3382.

[26]    J.Y. Zhang, G. Liu, R.H. Wang, J. Li, J. Sun and E. Ma, *Double-inverse grain size dependence of deformation twinning in nanocrystalline Cu.* Physical Review B, 2010. **81**(17).

[27]     X. Guo, W. Liang and M. Zhou, *Mechanism for the Pseudoelastic Behavior of FCC Shape Memory Nanowires.* Experimental Mechanics, 2009. **49**(2): p. 183-190.

[28]     B. Hyde, H.D. Espinosa and D. Farkas, *An atomistic investigation of elastic and plastic properties of Au nanowires.* Jom, 2005. **57**(9): p. 62-66.

[29]     C.J. Ji and H.S. Park, *The coupled effects of geometry and surface orientation on the mechanical properties of metal nanowires.* Nanotechnology, 2007. **18**(30).

[30]     A.M. Leach, M. McDowell and K. Gall, *Deformation of top-down and bottom-up silver nanowires.* Advanced Functional Materials, 2007. **17**(1): p. 43-53.

[31]     H.S. Park, K. Gall and J.A. Zimmerman, *Deformation of FCC nanowires by twinning and slip.* Journal of the Mechanics and Physics of Solids, 2006. **54**(9): p. 1862-1881.

[32]     F. Sansoz, H.C. Huang and D.H. Warner, *An atomistic perspective on twinning phenomena in nano-enhanced fcc metals.* JOM, 2008. **60**(9): p. 79-84.

[33]     J.H. SeoY. YooN.Y. ParkS.W. YoonH. LeeS. HanS.W. LeeT.Y. SeongS.C. LeeK.B. LeeP.R. ChaH.S. ParkB. KimJ.P. Ahn, *Superplastic Deformation of Defect-Free Au Nanowires via Coherent Twin Propagation.* Nano Letters, 2011. **11**(8): p. 3499-3502.

[34]     H. Zheng, A.J. Cao, C.R. Weinberger, J.Y. Huang, K. Du, J.B. Wang, Y.Y. Ma, Y.N. Xia and S.X. Mao, *Discrete plasticity in sub-10-nm-sized gold crystals.* Nature Communications, 2010. **1**.

[35]   H. Van Swygenhoven, P.M. Derlet and A.G. Froseth, *Stacking fault energies and slip in nanocrystalline metals.* Nature Materials, 2004. **3**(6): p. 399-403.

[36]   Z.H. Jin, S.T. Dunham, H. Gleiter, H. Hahn and P. Gumbsch, *A universal scaling of planar fault energy barriers in face-centered cubic metals.* Scripta Materialia, 2011. **64**(7): p. 605-608.

[37]   E.B. Tadmor and S. Hai, *A Peierls criterion for the onset of deformation twinning at a crack tip.* Journal of the Mechanics and Physics of Solids, 2003. **51**(5): p. 765-793.

[38]   D.H. Warner, W.A. Curtin and S. Qu, *Rate dependence of crack-tip processes predicts twinning trends in f.c.c. metals.* Nature Materials, 2007. **6**(11): p. 876-881.

[39]   J.R. Rice, *Dislocation Nucleation from a Crack Tip - an Analysis Based on the Peierls Concept.* Journal of the Mechanics and Physics of Solids, 1992. **40**(2): p. 239-271.

[40]   P. Gumbsch, *Elementary mechanisms of brittle and semi-brittle fracture.* Journal De Physique Iv, 2003. **106**: p. 3-12.

[41]   R. Danzer, T. Lube, P. Supancic and R. Damani, *Fracture of ceramics.* Advanced Engineering Materials, 2008. **10**(4): p. 275-298.

[42]   A.A. Griffith, *The Phenomenon of Rupture and Flow in Solids.* Phil. Trans. Royal Soc. London, 1920. **7**(A221): p. 163.

[43]   D. Munz and T. Fett, *Ceramics : mechanical properties, failure behaviour, materials selection.* 1999, Berlin ; London: Springer. x, 298p.

[44]   W. Weibull, *A Statistical Theory Of The Strength Of Materials.* Ingeniörsvetenskapsakademiens Handlingar, 1939. **151**.

[45]   C.S. John, *The brittle-to-ductile transition in pre-cleaved silicon single crystals.* Philosophical Magazine a-Physics of Condensed Matter Structure Defects and Mechanical Properties, 1975. **32**(6): p. 1193-1212.

[46]   M.S. Duesbery and B. Joos, *Dislocation motion in silicon: The shuffle-glide controversy.* Philosophical Magazine Letters, 1996. **74**(4): p. 253-258.

[47]   C.Z. Wang, J. Li, K.M. Ho and S. Yip, *Undissociated screw dislocation in Si: Glide or shuffle set?* Applied Physics Letters, 2006. **89**(5).

[48]   J. Ro\sler and H. Harders, *Mechanical behaviour of engineering materials : metals, ceramics, polymers, and composites.* 2007, Berlin ; New York: Springer. xv, 534 p.

[49]   M.T. Sprackling, *The plastic deformation of simple ionic crystals.* 1976, London: Academic Press. ix, 242p.

[50]   D. Hull and D.J. Bacon, *Introduction to dislocations.* 4th ed., by D. Hull and D.J. Bacon. ed. 2001, Oxford: Butterworth Heinemann. vii, 242 p.

[51]   W.S. Rothwell and R.G. Greenler, *Annealing and Yield Stress of Nacl Single Crystals.* Journal of the American Ceramic Society, 1964. **47**(11): p. 585-587.

[52]   A.F. Joffé, M.V. Kirpitcheva and M.A. Lewitzky, *Deformation und Festigkeit der Kristalle.* Zeitschrift für Physik, 1924. **22**(1): p. 286-302.

[53]   A.E. Gorum, E.R. Parker and J.A. Pask, *Effect of Surface Conditions on Room-Temperature Ductility of Ionic Crystals.* Journal of the American Ceramic Society, 1958. **41**(5): p. 161-164.

[54]    A.F. Joffé and M.V. Kirpitcheva, *XVIII. Röntgenograms of strained crystals.* Philosophical Magazine a-Physics of Condensed Matter Structure Defects and Mechanical Properties, 1922. **6**(43): p. 204-206.

[55]    P.L. Pratt, *Similar Glide Processes in Ionic and Metallic Crystals.* Acta Metallurgica, 1953. **1**(1): p. 103-104.

[56]    J.J. Gilman, C. Knudsen and W.P. Walsh, *Cleavage Cracks and Dislocations in Lif-Crystals.* Journal of Applied Physics, 1958. **29**(4): p. 601-607.

[57]    S.J. Burns and W.W. Webb, *Plastic Deformation during Cleavage of Lif.* Transactions of the Metallurgical Society of Aime, 1966. **236**(8): p. 1165-&.

[58]    A.J. Forty, *The Generation of Dislocations during Cleavage.* Proceedings of the Royal Society of London Series a-Mathematical and Physical Sciences, 1957. **242**(1230): p. 392-399.

[59]    G.T. Murray, *Brittle-Ductile Transition Temperatures in Ionic Crystals.* Journal of the American Ceramic Society, 1960. **43**(6): p. 330-334.

[60]    J. Hesse, *Die Plastische Verformung Von Natriumchlorid.* Physica Status Solidi, 1965. **9**(1): p. 209-&.

[61]    F.T. Wimmer, M.E. Fine and W. Kobes, *Tensile Properties of Nacl-Banr Solid Solution Single Crystals.* Journal of Applied Physics, 1963. **34**(6): p. 1775-&.

[62]    N. Gane, *Compressive Strength of Sub-Micrometer Diameter Magnesium-Oxide Crystals.* Philosophical Magazine, 1972. **25**(1): p. 25-&.

[63]    W.C. Oliver and G.M. Pharr, *An Improved Technique for Determining Hardness and Elastic-Modulus Using Load and*

*Displacement Sensing Indentation Experiments.* Journal of Materials Research, 1992. **7**(6): p. 1564-1583.

[64]  W.C. Oliver and G.M. Pharr, *Measurement of hardness and elastic modulus by instrumented indentation: Advances in understanding and refinements to methodology.* Journal of Materials Research, 2004. **19**(1): p. 3-20.

[65]  F.R. Brotzen, *Mechanical Testing of Thin-Films.* International Materials Reviews, 1994. **39**(1): p. 24-45.

[66]  D.A. Hardwick, *The Mechanical-Properties of Thin-Films - a Review.* Thin Solid Films, 1987. **154**(1-2): p. 109-124.

[67]  O. Kraft and C.A. Volkert, *Mechanical testing of thin films and small structures.* Advanced Engineering Materials, 2001. **3**(3): p. 99-110.

[68]  W.D. Nix, *Mechanical-Properties of Thin-Films.* Metallurgical Transactions a-Physical Metallurgy and Materials Science, 1989. **20**(11): p. 2217-2245.

[69]  M. Ohring, *The materials science of thin films.* 2nd ed. ed. 2001, San Diego, Calif. ; London: Academic. 800p.

[70]  L.B. Freund and S. Suresh, *Thin film materials : stress, defect formation and surface evolution.* 2003, Cambridge: Cambridge University Press. 850 p.

[71]  W.D. Beams, *Structure and properties of thin films*, Wiley, 1959, pp. 183-192.

[72]  J.J. Vlassak and W.D. Nix, *A New Bulge Test Technique for the Determination of Young Modulus and Poisson Ratio of Thin-Films.* Journal of Materials Research, 1992. **7**(12): p. 3242-3249.

[73]   G.G. Stoney, *The tension of metallic films deposited by electrolysis.* Proceedings of the Royal Society of London Series a-Mathematical and Physical Sciences, 1909. **82**(553): p. 172-175.

[74]   J.A. Ruud, D. Josell, F. Spaepen and A.L. Greer, *A New Method for Tensile Testing of Thin-Films.* Journal of Materials Research, 1993. **8**(1): p. 112-117.

[75]   D.T. Read and J.W. Dally, *A New Method for Measuring the Strength and Ductility of Thin-Films.* Journal of Materials Research, 1993. **8**(7): p. 1542-1549.

[76]   M.A. Haque and M.T.A. Saif, *In-situ tensile testing of nano-scale specimens in SEM and TEM.* Experimental Mechanics, 2002. **42**(1): p. 123-128.

[77]   M. Hommel and O. Kraft, *Deformation behavior of thin copper films on deformable substrates.* Acta Materialia, 2001. **49**(19): p. 3935-3947.

[78]   L. Schadler and I.C. Noyan, *Quantitative Measurement of the Stress Transfer-Function in Nickel/Polyimide Thin-Film Copper Thin-Film Structures.* Applied Physics Letters, 1995. **66**(1): p. 22-24.

[79]   F. Macionczyk, W. Bruckner, W. Pitschte and G. Reiss, *Young's modulus and tensile strength of CuNi(Mn) thin films on polyimide foils by tensile testing.* Journal of Materials Research, 1998. **13**(10): p. 2852-2858.

[80]   L.S. Schadler and I.C. Noyan, *Experimental-Determination of the Strain Transfer across a Flexible Intermediate Layer in Thin-Film Structures.* Journal of Materials Science Letters, 1992. **11**(15): p. 1067-1069.

[81]   M.D. Uchic, D.M. Dimiduk, J.N. Florando and W.D. Nix, *Exploring specimen size effects in plastic deformation of Ni-3(Al, Ta).* Defect

Properties and Related Phenomena in Intermetallic Alloys, 2003. **753**: p. 27-32.

[82]  D. Kiener, C. Motz, M. Rester, M. Jenko and G. Dehm, *FIB damage of Cu and possible consequences for miniaturized mechanical tests.* Materials Science and Engineering a-Structural Materials Properties Microstructure and Processing, 2007. **459**(1-2): p. 262-272.

[83]  Z.W. Shan, R.K. Mishra, S.A.S. Asif, O.L. Warren and A.M. Minor, *Mechanical annealing and source-limited deformation in submicrometre-diameter Ni crystals.* Nature Materials, 2008. **7**(2): p. 115-119.

[84]  S. Shim, H. Bei, M.K. Miller, G.M. Pharr and E.P. George, *Effects of focused ion beam milling on the compressive behavior of directionally solidified micropillars and the nanoindentation response of an electropolished surface.* Acta Materialia, 2009. **57**(2): p. 503-510.

[85]  H. Bei, S. Shim, E.P. George, M.K. Miller, E.G. Herbert and G.M. Pharr, *Compressive strengths of molybdenum alloy micro-pillars prepared using a new technique.* Scripta Materialia, 2007. **57**(5): p. 397-400.

[86]  D.S. Gianola, A. Sedlmayr, R. Monig, C.A. Volkert, R.C. Major, E. Cyrankowski, S.A.S. Asif, O.L. Warren and O. Kraft, *In situ nanomechanical testing in focused ion beam and scanning electron microscopes.* Review of Scientific Instruments, 2011. **82**(6).

[87]  S.S. Brenner, *Tensile Strength of Whiskers.* Journal of Applied Physics, 1956. **27**(12): p. 1484-1491.

[88]  S.S. Brenner, *Plastic Deformation of Copper and Silver Whiskers.* Journal of Applied Physics, 1957. **28**(9): p. 1023-1026.

[89]    D.M. Marsh, *Micro-Tensile Testing Machine.* Journal of Scientific Instruments, 1961. **38**(6): p. 229-&.

[90]    J.J. Petrovic, J.V. Milewski, D.L. Rohr and F.D. Gac, *Tensile Mechanical-Properties of Sic Whiskers.* Journal of Materials Science, 1985. **20**(4): p. 1167-1177.

[91]    H.D. Espinosa, B.C. Prorok and M. Fischer, *A methodology for determining mechanical properties of freestanding thin films and MEMS materials.* Journal of the Mechanics and Physics of Solids, 2003. **51**(1): p. 47-67.

[92]    M.A. Haque and M.T.A. Saif, *Microscale materials testing using MEMS actuators.* Journal of Microelectromechanical Systems, 2001. **10**(1): p. 146-152.

[93]    D. Kiener, W. Grosinger, G. Dehm and R. Pippan, *A further step towards an understanding of size-dependent crystal plasticity: In situ tension experiments of miniaturized single-crystal copper samples.* Acta Materialia, 2008. **56**(3): p. 580-592.

[94]    M. Naraghi, T. Ozkan, I. Chasiotis, S.S. Hazra and M.P. de Boer, *MEMS platform for on-chip nanomechanical experiments with strong and highly ductile nanofibers.* Journal of Micromechanics and Microengineering, 2010. **20**(12).

[95]    S. Orso, U.G.K. Wegst, C. Eberl and E. Arzt, *Micrometer-scale tensile testing of biological attachment devices.* Advanced Materials, 2006. **18**(7): p. 874-+.

[96]    K.A. Rzepiejewska-Malyska, G. Buerki, J. Michler, R.C. Major, E. Cyrankowski, S.A.S. Asif and O.L. Warren, *In situ mechanical observations during nanoindentation inside a high-resolution scanning electron microscope.* Journal of Materials Research, 2008. **23**(7): p. 1973-1979.

[97]  W.N. Sharpe, B. Yuan and R.L. Edwards, *A new technique for measuring the mechanical properties of thin films.* Journal of Microelectromechanical Systems, 1997. **6**(3): p. 193-199.

[98]  D.F. Zhang, J.M. Breguet, R. Clavel, V. Sivakov, S. Christiansen and J. Michler, *Electron Microscopy Mechanical Testing of Silicon Nanowires Using Electrostatically Actuated Tensile Stages.* Journal of Microelectromechanical Systems, 2010. **19**(3): p. 663-674.

[99]  M.F. Yu, M.J. Dyer, G.D. Skidmore, H.W. Rohrs, X.K. Lu, K.D. Ausman, J.R. Von Ehr and R.S. Ruoff, *Three-dimensional manipulation of carbon nanotubes under a scanning electron microscope.* Nanotechnology, 1999. **10**(3): p. 244-252.

[100]  M.F. Yu, O. Lourie, M.J. Dyer, K. Moloni, T.F. Kelly and R.S. Ruoff, *Strength and breaking mechanism of multiwalled carbon nanotubes under tensile load.* Science, 2000. **287**(5453): p. 637-640.

[101]  H.D. Espinosa, Y. Zhu and N. Moldovan, *Design and operation of a MEMS-based material testing system for nanomechanical characterization.* Journal of Microelectromechanical Systems, 2007. **16**(5): p. 1219-1231.

[102]  T. Zhu and J. Li, *Ultra-strength materials.* Progress in Materials Science, 2010. **55**(7): p. 710-757.

[103]  E. Arzt, *Overview no. 130 - Size effects in materials due to microstructural and dimensional constraints: A comparative review.* Acta Materialia, 1998. **46**(16): p. 5611-5626.

[104]  O. Kraft, P.A. Gruber, R. Monig and D. Weygand, *Plasticity in Confined Dimensions.* Annual Review of Materials Research, Vol 40, 2010. **40**: p. 293-317.

[105]  T.T. Zhu, A.J. Bushby and D.J. Dunstan, *Materials mechanical size effects: a review.* Materials Technology, 2008. **23**(4): p. 193-209.

[106] E.O. Hall, *The Deformation and Ageing of Mild Steel .3. Discussion of Results.* Proceedings of the Physical Society of London Section B, 1951. **64**(381): p. 747-753.

[107] N.J. Petch, *The Cleavage Strength of Polycrystals.* Journal of the Iron and Steel Institute, 1953. **174**(1): p. 25-28.

[108] G.F. Taylor, *A Method of Drawing Metallic Filaments and a Discussion of their Properties and Uses.* Physical Review, 1924. **23**(5): p. 655-660.

[109] G. Richter, K. Hillerich, D.S. Gianola, R. Monig, O. Kraft and C.A. Volkert, *Ultrahigh Strength Single Crystalline Nanowhiskers Grown by Physical Vapor Deposition.* Nano Letters, 2009. **9**(8): p. 3048-3052.

[110] P.A. Gruber, C. Solenthaler, E. Arzt and R. Spolenak, *Strong single-crystalline Au films tested by a new synchrotron technique.* Acta Materialia, 2008. **56**(8): p. 1876-1889.

[111] C.P. Frick, B.G. Clark, S. Orso, A.S. Schneider and E. Arzt, *Size effect on strength and strain hardening of small-scale [111] nickel compression pillars.* Materials Science and Engineering a-Structural Materials Properties Microstructure and Processing, 2008. **489**(1-2): p. 319-329.

[112] J.R. Greer, W.C. Oliver and W.D. Nix, *Size dependence of mechanical properties of gold at the micron scale in the absence of strain gradients.* Acta Materialia, 2005. **53**(6): p. 1821-1830.

[113] K.S. Ng and A.H.W. Ngan, *Stochastic nature of plasticity of aluminum micro-pillars.* Acta Materialia, 2008. **56**(8): p. 1712-1720.

[114]  C.A. Volkert and E.T. Lilleodden, *Size effects in the deformation of sub-micron Au columns.* Philosophical Magazine, 2006. **86**(33-35): p. 5567-5579.

[115]  D.M. Dimiduk, C. Woodward, R. LeSar and M.D. Uchic, *Scale-free intermittent flow in crystal plasticity.* Science, 2006. **312**(5777): p. 1188-1190.

[116]  T.A. Parthasarathy, S.I. Rao, D.M. Dimiduk, M.D. Uchic and D.R. Trinkle, *Contribution to size effect of yield strength from the stochastics of dislocation source lengths in finite samples.* Scripta Materialia, 2007. **56**(4): p. 313-316.

[117]  S.I. Rao, D.M. Dimiduk, T.A. Parthasarathy, M.D. Uchic, M. Tang and C. Woodward, *Athermal mechanisms of size-dependent crystal flow gleaned from three-dimensional discrete dislocation simulations.* Acta Materialia, 2008. **56**(13): p. 3245-3259.

[118]  S.I. Rao, D.M. Dimiduk, M. Tang, T.A. Parthasarathy, M.D. Uchic and C. Woodward, *Estimating the strength of single-ended dislocation sources in micron-sized single crystals.* Philosophical Magazine, 2007. **87**(30): p. 4777-4794.

[119]  F. Mompiou, M. Legros, A. Sedlmayr, D.S. Gianola, D. Caillard and O. Kraft, *Source-based strengthening of sub-micrometer Al fibers.* Acta Materialia, 2012. **60**: p. 977-983.

[120]  D. Kiener, C. Motz, T. Schoberl, M. Jenko and G. Dehm, *Determination of mechanical properties of copper at the micron scale.* Advanced Engineering Materials, 2006. **8**(11): p. 1119-1125.

[121]  H. Bei, S. Shim, G.M. Pharr and E.P. George, *Effects of pre-strain on the compressive stress-strain response of Mo-alloy single-crystal micropillars.* Acta Materialia, 2008. **56**(17): p. 4762-4770.

[122] O.M. Jadaan, N.N. Nemeth, J. Bagdahn and W.N. Sharpe, *Probabilistic Weibull behavior and mechanical properties of MEMS brittle materials.* Journal of Materials Science, 2003. **38**(20): p. 4087-4113.

[123] R.L. Eisner, *Tensile Tests on Silicon Whiskers.* Acta Metallurgica, 1955. **3**(4): p. 414-415.

[124] G.L. Pearson, W.T. Read and W.L. Feldmann, *Deformation and Fracture of Small Silicon Crystals.* Acta Metallurgica, 1957. **5**(4): p. 181-191.

[125] J. Cook, C.C. Evans, J.E. Gordon and D.M. Marsh, *Mechanism for Control of Crack Propagation in All-Brittle Systems.* Proceedings of the Royal Society of London Series a-Mathematical and Physical Sciences, 1964. **282**(1390): p. 508-+.

[126] S.M. Hu, *Critical Stress in Silicon Brittle-Fracture, and Effect of Ion-Implantation and Other Surface Treatments.* Journal of Applied Physics, 1982. **53**(5): p. 3576-3580.

[127] T. Namazu, Y. Isono and T. Tanaka, *Evaluation of size effect on mechanical properties of single crystal silicon by nanoscale bending test using AFM.* Journal of Microelectromechanical Systems, 2000. **9**(4): p. 450-459.

[128] H.S. Park, W. Cai, H.D. Espinosa and H.C. Huang, *Mechanics of Crystalline Nanowires (vol 34, pg 178, 2009).* Mrs Bulletin, 2009. **34**(4): p. 232-232.

[129] M.J. Gordon, T. Baron, F. Dhalluin, P. Gentile and P. Ferret, *Size Effects in Mechanical Deformation and Fracture of Cantilevered Silicon Nanowires.* Nano Letters, 2009. **9**(2): p. 525-529.

[130]  Y. Zhu, F. Xu, Q.Q. Qin, W.Y. Fung and W. Lu, *Mechanical Properties of Vapor-Liquid-Solid Synthesized Silicon Nanowires.* Nano Letters, 2009. **9**(11): p. 3934-3939.

[131]  B. Lee and R.E. Rudd, *First-principles study of the Young's modulus of Si < 001 > nanowires.* Physical Review B, 2007. **75**(4).

[132]  H.W. Shim, L.G. Zhou, H.C. Huang and T.S. Cale, *Nanoplate elasticity under surface reconstruction.* Applied Physics Letters, 2005. **86**(15).

[133]  K. Kang and W. Cai, *Brittle and ductile fracture of semiconductor nanowires - molecular dynamics simulations.* Philosophical Magazine, 2007. **87**(14-15): p. 2169-2189.

[134]  X.D. Han, K. Zheng, Y.F. Zhang, X.N. Zhang, Z. Zhang and Z.L. Wang, *Low-temperature in situ large-strain plasticity of silicon nanowires.* Advanced Materials, 2007. **19**(16): p. 2112-+.

[135]  Z. Gyulai, *Festigkeits Und Plastizitatseigenschaften Von Nacl-Nadelkristallen.* Zeitschrift Fur Physik, 1954. **138**(3-4): p. 317-321.

[136]  R.S. Wagner and W.C. Ellis, *Vapor-Liquid-Solid Mechanism of Single Crystal Growth ( New Method Growth Catalysis from Impurity Whisker Epitaxial + Large Crystals Si E ).* Applied Physics Letters, 1964. **4**(5): p. 89-&.

[137]  E.I. Givargizov, *Fundamental Aspects of Vls Growth.* Journal of Crystal Growth, 1975. **31**(Dec): p. 20-30.

[138]  K.A. Dick, *A review of nanowire growth promoted by alloys and non-alloying elements with emphasis on Au-assisted III-V nanowires.* Progress in Crystal Growth and Characterization of Materials, 2008. **54**(3-4): p. 138-173.

[139]  V. Schmidt, J.V. Wittemann, S. Senz and U. Gosele, *Silicon Nanowires: A Review on Aspects of their Growth and their*

*Electrical Properties.* Advanced Materials, 2009. **21**(25-26): p. 2681-2702.

[140] Y. Cui, L.J. Lauhon, M.S. Gudiksen, J.F. Wang and C.M. Lieber, *Diameter-controlled synthesis of single-crystal silicon nanowires.* Applied Physics Letters, 2001. **78**(15): p. 2214-2216.

[141] W. Lu and C.M. Lieber, *Semiconductor nanowires.* Journal of Physics D-Applied Physics, 2006. **39**(21): p. R387-R406.

[142] A.M. Morales and C.M. Lieber, *Rational synthesis of silicon nanowires.* Abstracts of Papers of the American Chemical Society, 1997. **213**: p. 651-INOR.

[143] Y. Wu, Y. Cui, L. Huynh, C.J. Barrelet, D.C. Bell and C.M. Lieber, *Controlled growth and structures of molecular-scale silicon nanowires.* Nano Letters, 2004. **4**(3): p. 433-436.

[144] O.M. Nayfeh, D.A. Antoniadis, S. Boles, C. Ho and C.V. Thompson, *Formation of Single Tiers of Bridging Silicon Nanowires for Transistor Applications Using Vapor-Liquid-Solid Growth from Short Silicon-on-Insulator Sidewalls.* Small, 2009. **5**(21): p. 2440-2444.

[145] X. Li and P.W. Bohn, *Metal-assisted chemical etching in HF/H(2)O(2) produces porous silicon.* Applied Physics Letters, 2000. **77**(16): p. 2572-2574.

[146] Z.P. Huang, X.X. Zhang, M. Reiche, L.F. Liu, W. Lee, T. Shimizu, S. Senz and U. Gosele, *Extended arrays of vertically aligned sub-10 nm diameter [100] Si nanowires by metal-assisted chemical etching.* Nano Letters, 2008. **8**(9): p. 3046-3051.

[147] C.L. Cheung, R.J. Nikolic, C.E. Reinhardt and T.F. Wang, *Fabrication of nanopillars by nanosphere lithography.* Nanotechnology, 2006. **17**(5): p. 1339-1343.

[148]  S.W. Chang, V.P. Chuang, S.T. Boles and C.V. Thompson, *Metal-Catalyzed Etching of Vertically Aligned Polysilicon and Amorphous Silicon Nanowire Arrays by Etching Direction Confinement.* Advanced Functional Materials, 2010. **20**(24): p. 4364-4370.

[149]  S.W. Chang, V.P. Chuang, S.T. Boles, C.A. Ross and C.V. Thompson, *Densely Packed Arrays of Ultra-High-Aspect-Ratio Silicon Nanowires Fabricated using Block-Copolymer Lithography and Metal-Assisted Etching.* Advanced Functional Materials, 2009. **19**(15): p. 2495-2500.

[150]  B. Bhushan, A.V. Kulkarni, W. Bonin and J.T. Wyrobek, *Nanoindentation and picoindentation measurements using a capacitive transducer system in atomic force microscopy.* Philosophical Magazine a-Physics of Condensed Matter Structure Defects and Mechanical Properties, 1996. **74**(5): p. 1117-1128.

[151]  cited 03.11.2011; Available from: http://www.femtotools.com/index.php?id=51.

[152]  Y. Sun, S.N. Fry, D.P. Potasek, D.J. Bell and B.J. Nelson, *Characterizing fruit fly flight behavior using a microforce sensor with a new comb-drive configuration.* Journal of Microelectromechanical Systems, 2005. **14**(1): p. 4-11.

[153]  D.S. Gianola and C. Eberl, *Micro- and nanoscale tensile testing of materials.* JOM, 2009. **61**(3): p. 24-35.

[154]  C. Eberl, D.S. Gianola and R. Thompson, *Digital Image Correlation and Tracking.* Matlab Central, 2006: p. File ID: #12413.

[155]  M.A. Sutton, J.-J. Orteu and H.W. Schreier, *Image correlation for shape, motion and deformation measurements : basic concepts, theory and applications.* 2009, New York: Springer. xx, 321 p.

[156] H.A. Bruck, S.R. Mcneill, M.A. Sutton and W.H. Peters, *Digital Image Correlation Using Newton-Raphson Method of Partial-Differential Correction.* Experimental Mechanics, 1989. **29**(3): p. 261-267.

[157] V. Gopal, E.A. Stach, V.R. Radmilovic and I.A. Mowat, *Metal delocalization and surface decoration in direct-write nanolithography by electron beam induced deposition.* Applied Physics Letters, 2004. **85**(1): p. 49-51.

[158] I. Utke, V. Friedli, S. Fahlbusch, S. Hoffmann, P. Hoffmann and J. Michler, *Tensile strengths of metal-containing joints fabricated by focused electron beam induced deposition.* Advanced Engineering Materials, 2006. **8**(3): p. 155-157.

[159] K.E. Johanns, A. Sedlmayr, P.S. Phani, R. Monig, O. Kraft, E.P. George and G.M. Pharr, *In-situ tensile testing of single-crystal molybdenum-alloy fibers with various dislocation densities in a scanning electron microscope.* Journal of Materials Research, 2012. **27**(3): p. 508-520.

[160] P.S. Phani, K.E. Johanns, G. Duscher, A. Gali, E.P. George and G.M. Pharr, *Scanning transmission electron microscope observations of defects in as-grown and pre-strained Mo alloy fibers.* Acta Materialia, 2011. **59**(5): p. 2172-2179.

[161] J.P. Pouget, H. Launois, J.P. Dhaenens, P. Merenda and T.M. Rice, *Electron Localization Induced by Uniaxial Stress in Pure Vo2.* Physical Review Letters, 1975. **35**(13): p. 873-875.

[162] H. Guo, K. Chen, Y. Oh, K. Wang, C. Dejoie, S.A.S. Asif, O.L. Warren, Z.W. Shan, J. Wu and A.M. Minor, *Mechanics and Dynamics of the Strain-Induced M1-M2 Structural Phase Transition in Individual VO(2) Nanowires.* Nano Letters, 2011. **11**(8): p. 3207-3213.

[163] Y. Cheng, T.L. Wong, K.M. Ho and N. Wang, *The structure and growth mechanism of VO(2) nanowires.* Journal of Crystal Growth, 2009. **311**(6): p. 1571-1575.

[164] V.N. Andreev, F.A. Chudnovskii, A.V. Petrov and E.I. Terukov, *Thermal-Conductivity of Vo2,V3o5, and V2o3.* Physica Status Solidi a-Applied Research, 1978. **48**(2): p. K153-K156.

[165] B. Audoly and S. Neukirch, *Fragmentation of rods by cascading cracks: Why spaghetti does not break in half.* Physical Review Letters, 2005. **95**(9).

[166] M.S. Steighner, L.P. Snedeker, B.L. Boyce, K. Gall, D.C. Miller and C.L. Muhlstein, *Dependence on diameter and growth direction of apparent strain to failure of Si nanowires.* Journal of Applied Physics, 2011. **109**(3).

[167] S. Hoffmann, I. Utke, B. Moser, J. Michler, S.H. Christiansen, V. Schmidt, S. Senz, P. Werner, U. Gosele and C. Ballif, *Measurement of the bending strength of vapor-liquid-solid grown silicon nanowires.* Nano Letters, 2006. **6**(4): p. 622-625.

[168] D. Kaufmann, *unpublished results.*

[169] T. TsuchiyaM. HirataN. ChibaR. UdoY. YoshitomiT. AndoK. SatoK. TakashimaY. HigoY. SaotomeH. OgawaK. Ozaki, *Cross comparison of thin-film tensile-testing methods examined using single-crystal silicon, polysilicon, nickel, and titanium films.* Journal of Microelectromechanical Systems, 2005. **14**(5): p. 1178-1186.

[170] S. Boles, *private communication.*

[171] R.G. Bryans and D.W.F. James, *The tensile strength of silicon whiskers grown by the vapour liquid solid technique.* Micron, 1970. **2**: p. 89-100.

[172] N. Macmillan and A. Kelly, *The Mechanical Properties of Perfect Crystals. I. The Ideal Strength.* Proceedings of the Royal Society a-Mathematical Physical and Engineering Sciences, 1972. **330**(1582): p. 291-308.

[173] N.W. Moore, J.H. Luo, J.Y. Huang, S.X. Mao and J.E. Houston, *Superplastic Nanowires Pulled from the Surface of Common Salt.* Nano Letters, 2009. **9**(6): p. 2295-2299.

[174] H.C. Allen, M.L. Mecartney and J.C. Hemminger, *Minimizing transmission electron microscopy beam damage during the study of surface reactions on sodium chloride.* Microscopy and Microanalysis, 1998. **4**(1): p. 23-33.

[175] G. Richter, *private communication.*

[176] S. Mendelson, *Dislocations and Plastic Flow in NaCl Single Crystals.* Journal of Applied Physics, 1962. **33**(7): p. 2175-&.

[177] S.S. Brenner and C.R. Morelock, *Cross Sectioning of Whiskers.* Review of Scientific Instruments, 1957. **28**(8): p. 652-653.

[178] A. Sedlmayr, E. Bitzek, D.S. Gianola, G. Richter, R. Mönig and O. Kraft, *Existence of Two Twinning-Mediated Plastic Deformation Modes in Au Nanowhiskers.* submitted to Acta Materialia, 2012.

[179] I. Karaman, H. Sehitoglu, K. Gall, Y.I. Chumlyakov and H.J. Maier, *Deformation of single crystal Hadfield steel by twinning and slip.* Acta Materialia, 2000. **48**(6): p. 1345-1359.

[180] Z.H. Jin, P. Gumbsch, E. Ma, K. Albe, K. Lu, H. Hahn and H. Gleiter, *The interaction mechanism of screw dislocations with coherent twin boundaries in different face-centred cubic metals.* Scripta Materialia, 2006. **54**(6): p. 1163-1168.

[181] H.S. Park and J.A. Zimmerman, *Modeling inelasticity and failure in gold nanowires.* Physical Review B, 2005. **72**(5).

[182]  E. Bitzek, *private communication.*

[183]  S. Brinckmann, J.Y. Kim, A. Jennings and J.R. Greer, *Effects of Sample Geometry on the Uniaxial Tensile Stress State at the Nanoscale.* International Journal for Multiscale Computational Engineering, 2009. **7**(3): p. 187-194.

[184]  S.S. Brenner, *Strength of Gold Whiskers.* Journal of Applied Physics, 1959. **30**(2): p. 266-267.

# A   Appendices

## A.1   Transducer Setup

### A.1.1   First Transducer Setup Design

The first setup design describes the status of the tensile testing system when this work was started. An optical image of the whole assembly inside the vacuum chamber is displayed in Figure A.1.1. In this version the Kleindiek manipulator is mounted to the standard Kleindiek bridgemount which is nearly u-shaped and in turn can be clamped by screws to the tilting part of the SEM-stage, causing the manipulator to tilt with the stage. The tip is movable only within its own range of motion. The Hysitron transducer is attached to the top of the xyz Attocube stack via the standard dovetail joint, which allows for easy fixation and removal. The bottom of the Attocube stack is again mounted to a customized L-mount that facilitates full range of motion for the positioners. This L-mount is fixed to a block which in turn is clamped to one of the slotted holes of the bridgemount. Thus the entire testing equipment is situated on the bridgemount allowing the transducer to be driven in and out of the field of view. The sample is mounted to a conventional sample holder providing free xyz and rotational motion of the sample stage.

In Figure A.1.1B a 3D computer aided design (CAD) image is shown, highlighting the degrees of freedom. The transducer can be driven into the field of view by the x- and z-Attocube. It can further be brought to the eucentric height by motion of the y-Attocube. The sample then can be brought to the probe of the transducer by motion of the microscope stage. From this schematic the cantilever like design of the transducer mounted to the Attocube stack and the L-mount is apparent. This design can be highly susceptible to mechanical noise in the system where the long cantilever with the transducer-tip at its free end offers a geometry that may amplify the vibration of the system.

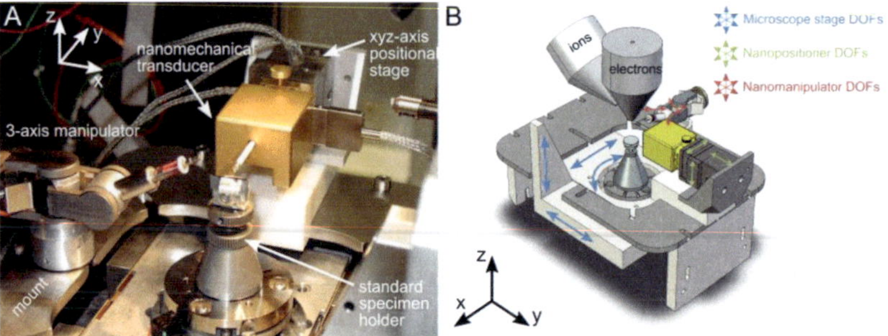


Figure A.1.1: First design of Hysitron setup inside the vacuum chamber of the dual-beam microscope. (A) shows an optical image of the assembly with the labeling of crucial parts. The transducer is mounted to the stack of three positioners through a dovetail joint where this cantilever like design makes it extremely susceptible to picking up mechanical noise. In a 3D-CAD drawing (B) the degrees of freedom for all moveable parts are highlighted.

Since with this design the vibration characteristics were much worse than specified by the company, noisefloor measurements were carried out at the KIT lab in Karlsruhe and the Hysitron facility in Minnesota, US for the sake of comparison. Measurements at KIT were accomplished by Dr. Daniel Gianola (KIT) and Dr. Ude Hangen (Hysitron). The data at Hysitron facility was acquired and processed by Ryan Major.

The transducer was tested in 3 different mounting configurations, also shown in Figure A.1.2:

- Transducer sitting in black holder, which was provided by Hysitron for storage and transport of the device. It was placed on an air table at KIT and on a Minus K anti-vibration table inside an acoustic enclosure at Hysitron.

- Setup on air table, where the transducer is fully assembled to the setup on an air table.

- Setup in FIB, where the transducer is mounted to the setup inside the vacuum chamber of the dual-beam.

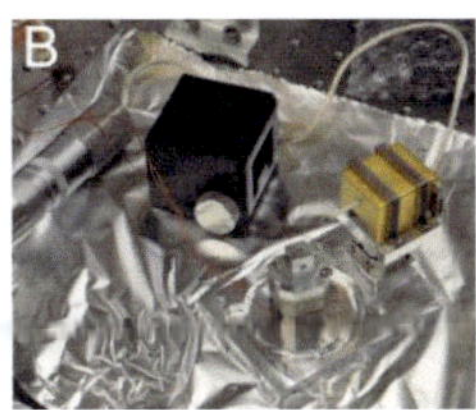

Figure A.1.2: Various mounting configurations for measurements of noise. In configuration 1 the transducer is kept inside the black transport holder provided by Hysitron (A). (B) shows the transducer mounted to the setup in air, whereas in (C) an optical image of the setup in vacuum can be seen.

No DC voltage was applied to the transducer, simply operating it as a displacement sensor. During operation in open-loop load control, zero force ramps of 10 seconds were employed at an acquisition rate of 1000 Hz. The root mean square (rms) noise floors and fast Fourier transform (FFT) spectra were taken for each set of data. The specified noise floor for a standard transducer of this design is 0.2 nm displacement 0.1 µN force standard deviation. In Figure A.1.3 the FFT spectra of the various cases are shown; the force and displacement noise floors are included.

The data taken with the transducer clamped in the black holder show the lowest noise floor of all measurements and flat spectra with no significant peaks, however the tests performed at Hysitron facility are still much lower. They are plotted in grey for comparison but can barely be seen due to the low noise. These conditions can be regarded as the "ideal case", providing a benchmark for comparison, whereas the increased noise at KIT can be attributed to the exposure of the device to ambient environment. Significantly higher values are seen for the whole setup placed on the air table. This data obviously indicates a problem related to the mounting of the transducer. The FFT spectrum shows a large peak around 100 Hz, indicating oscillation at the natural resonance frequency of the transducer. The third data set inside the vacuum chamber still shows high noise. This could indicate that there is some fundamental issue with assembling and mounting the system. The FFT shows no peak around 100 Hz, but a broad peak at 40 Hz. The source for this remains unknown, but the peak could be due to a poor anti-vibration system on the dual-beam microscope or even come from a cable or some other component in the system.

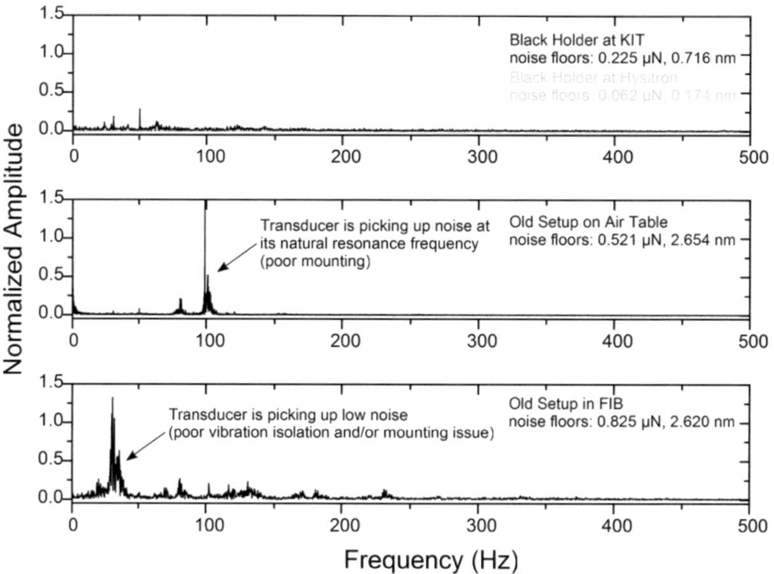


Figure A.1.3: FFT spectra of various testing configurations for the first setup. The large increase in noise when mounted to the setup indicates that the cantilever design is intensifying the vibrations of the transducer.

Overall it seems like the large increase in vibration derives from poor mounting of the system. The Hysitron transducer plus other components stacked build a large cantilever that enables mechanical vibration of the setup. The mechanical stiffness of the Attocube stack is rather low, so that together with the elongated horizontal geometry only low mechanical resonance frequencies can be achieved. Such a configuration is therefore more susceptible for picking up mechanical noise than a short and stiff system with a high resonance frequency. This could be combined with acoustic noise coming from the environment (building, vacuum pumps, etc.). The anti-vibration system of the FEI Nova Nanolab 200 is optimized for imaging but has to carry much higher loads in this case. It may be not perfect in damping out acoustic noise.

Since the noise floor is significantly higher with the transducer mounted to the setup, the first task is to change the design of the setup to a less compliant construction. In this cantilever design the component stiffness and assembly needs to be carefully addressed. With a stiffer design it should be possible to minimize the noise related to the mounting of the transducer. Furthermore, it should be made sure that cables hanging loose from the Attocube stages are mechanically fixed to exclude the possibility of noise from moving cables. There is also a need to electrically ground all stages and components properly. Ideally the noise floor should be the same in mounted stage assembly or if it is left isolated. Hence the goal is to get a configuration as good as the noise of the black holder measurement at KIT.

A comparison of force and displacement noise floors and the improvement with the design iteration are shown in Figure A.1.4.

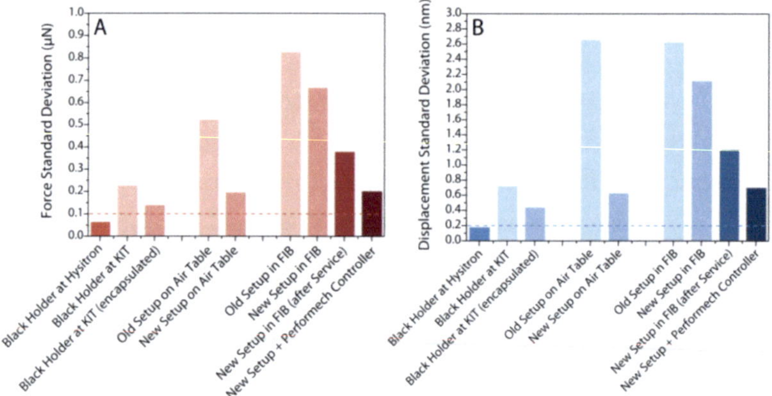

Figure A.1.4: Force and displacement noisefloors for the transducer mounted in different configurations. The new setup shows significant improvement in air. By employing new controller hardware including new software algorithms the noisefloor was further reduced, decreasing by 3.5 times. The dashed lines represent the standard reference values for this type of transducer.

## A.1.2  Peak Analysis

In order to elucidate and eliminate the origin of the remaining noise in the system, the peaks around the natural resonance frequency need to be checked carefully. If the region around 100 Hz is magnified there are three peaks visible (black curve in Figure A.1.5A). Hysitron Inc. supposed that the peak around 100 Hz could derive from electrical noise from the power line. Since they use a full-wave rectification to create direct current, the line frequency is doubled and would denote a frequency of 100 Hz in the system. To determine whether the peaks are mechanically or electrically driven, two approaches were taken. The resonance frequency $\omega_0$ of the transducer depends on the stiffness $c$ and the mass $m$ of the system (Equation A.1). Removal of the tip should shift the mechanical related peaks to slightly higher frequencies.

$$\omega_0 = \sqrt{\frac{c}{m}} \tag{A.1}$$

Figure A.1.5B shows the recorded spectrum without the tip. Even if the peak at 100.5 Hz is significantly reduced, only the peak at 102.3 Hz is shifted to a higher frequency. To confirm that only this peak derives from mechanical resonance the transducer was excited by bumping against the outside of the microscope vacuum chamber before measuring the noise. Figure A.1.5C displays huge amplitude at 102.3 Hz, however the other peaks are no more apparent.

All proposed origins of noise were addressed one-by-one in strategic measurements. The aim was to exclude variables and hence possible sources of noise from the overall system. Various configurations comprised changes in line source, FIB state, grounding, z-Attocube stage. Even if the mechanical peak was almost damped out in air, the electrical peak remained stable.

However during additional noise floor measurements a few weeks later (09/2009), the recorded noise floor almost halved. This fact is mostly caused by a substantial reduction of the 100.5 Hz electrical peak (see red curve in Figure A.1.5A). The setup configuration has not been changed since 08/2009 and so the origin of this eminent improvement is still not revealed. Most likely the enhanced noise floor derives from a FEI service done shortly before the last measurements, when an electronics board of the Nova Nanolab was replaced. The green curve in Figure A.1.5A shows further reduction of the noise by using the newly designed Performech controller with built-in Q-control.

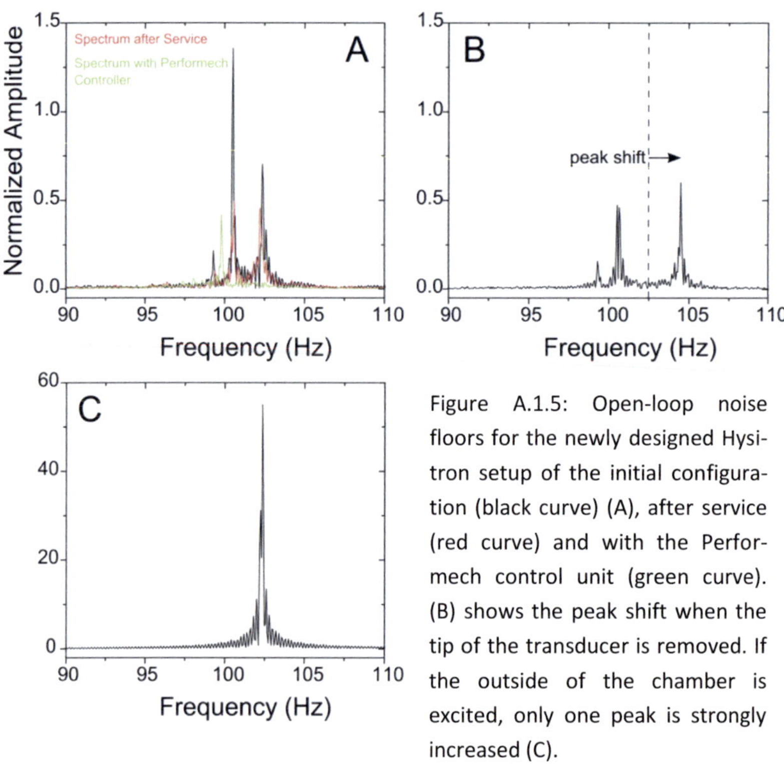



Figure A.1.5: Open-loop noise floors for the newly designed Hysitron setup of the initial configuration (black curve) (A), after service (red curve) and with the Performech control unit (green curve). (B) shows the peak shift when the tip of the transducer is removed. If the outside of the chamber is excited, only one peak is strongly increased (C).

## A.1.3 Strain-rate control of transducer

In order to track mechanical events like load-drops and to accurately determine the strain rate, the experiments are conducted in closed-loop displacement control. A PID feedback algorithm, where the parameters can be tuned accordingly prior to each experiment, is used for controlling the displacement. However, it has been observed in various experiments that upon a load-drop the controller is not able to maintain control. Load-drops are accompanied by a sudden release in elastic strain energy and result in a forward surge of the transducer. Due to the inertia of the tip this effect causes a certain amount of uncontrolled motion and extra plastic extension of the specimen. Depending on the elastic energy small (Figure A.1.6A) or large (Figure A.1.6B) amounts of uncontrolled displacement are obtained. When the PID regains control it realizes that it is far ahead of the target position and subsequently retracts until this position is reached, causing the sample to buckle. During the short period without control the strain-rate is suspected to significantly differ from the nominal strain-rate.

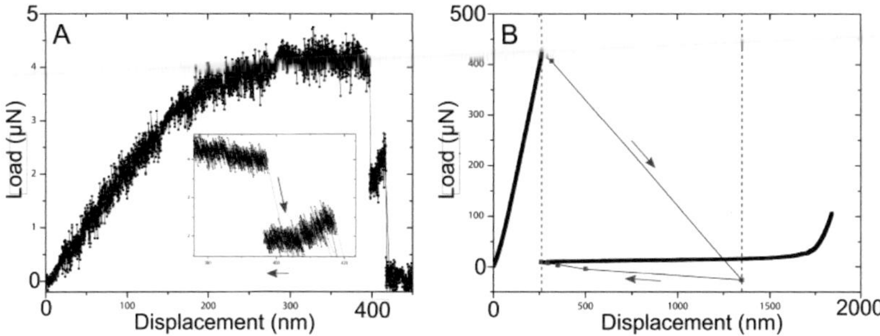

Figure A.1.6: Load-displacement curves for Au (A) and Mo (B). At the load-drop both curves show an uncontrolled forward surge of the transducer, which is retracted to the target position when the PID regained control. During this period the strain-rate may differ significantly from the nominal strain-rate.

## A.1.4  Lab VIEW Program for Tensile Testing

A Lab VIEW Program was written to incorporate the actuation parameters of the Attocube ANP Nanopositioners and the control of a Keithley 200 voltmeter for tensile and optional electrical testing. With this program the piezo crystals could be actuated remotely and the acquired dataset is concurrently stored in one single file. Figure A.1.7 shows the front panel (Figure A.1.7A) and the block diagram (Figure A.1.7B) of Lab VIEW VIs for the control of the Attocube positioners and the Keithley 200 voltmeter.

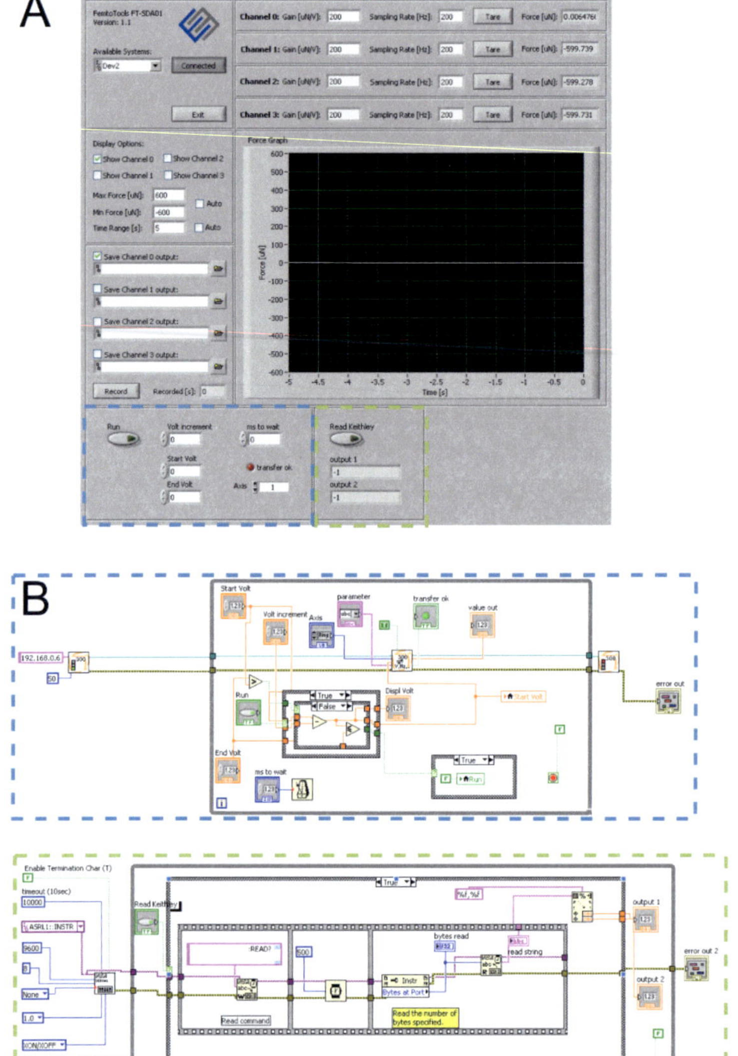

Figure A.1.7: LabVIEW program for the control of the Attocube Positioners and the Keithley 200 voltmeter.

## A.2    In situ Nanowire Battery Setup

Charging and discharging single nanowires inside a SEM is a non-trivial task. Many parameters have to be considered carefully to achieve successful testing in a vacuum environment. Like in any other LIB the cell needs to be assembled in the absence of air and humidity, which is usually done in a glovebox. Crucial to the success of the experiment is the transfer between the glovebox (usually Ar atmosphere) and the vacuum chamber of the microscope, where the cell is then opened under vacuum atmosphere. Although commercial systems are available, a dedicated transfer system was developed recently at IAM/KIT to assemble cells in a glovebox and test them *in situ* inside the SEM while concurrently observing the changes in the electrode material associated with lithiation and delithiation. The development and application of this method to particle based electrodes was performed within the Ph. D. work of Di Chen. Here it is used for nanowires. An optical image of the test system is shown in Figure A.2.1. It consists of two main parts and an opening mechanism. The bottom part is mounted on the microscope rotational stage. A conventional SEM stub can be equipped with battery equipment and fixed on top of this part. The upper part is basically a cap which is used to cover the cell by a thread and an O-ring in between. On top two screws are fixed. For opening, the whole device is moved to a fork which is mounted to the chamber wall. Once the two screws are placed in the fork continuous rotation of the microscope stage is used to open the cell. Using the same approach in the other direction the system can be closed under vacuum after testing.

For *in situ* lithiation of nanowires the arrangement of items on the SEM stub is shown in Figure A.2.1B and is as follows: A nanowire sample is mounted close to the side of the stub for decollating single specimens from the forest. A rod made of copper is pushed tightly into the stub. On top of the rod pure lithium is attached. Plastic deformation of the lithium helps to ensure a proper electrical connection. A 0.5 M solution of lithium-bis-

(trifluoromethanesulfonyl)imide (Li-TFSI lithium conductive salt) in buthylmethylpyrolidinium-TFSI (BMPyrr-TFSI, ionic liquid, provided by Ionic Liquid Technologies GmbH, Heilbronn, Germany) is used, which is trickled onto the Li. To avoid movements of the electrolyte it can be soaked into filter paper made of glass fibers or into a metallic mesh. For the storage of cycled specimens a standard TEM grid can be additionally mounted to the stub.

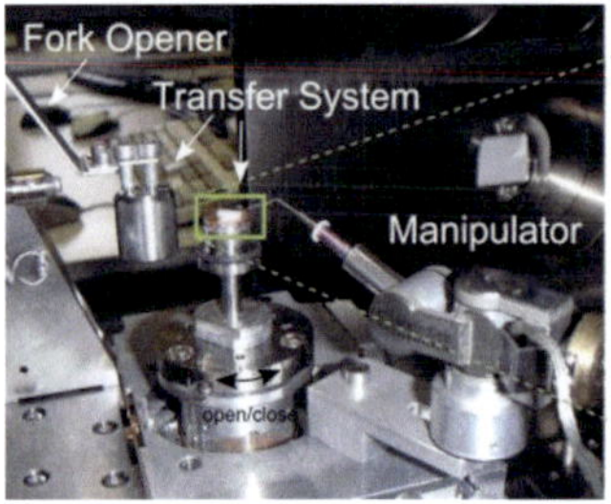

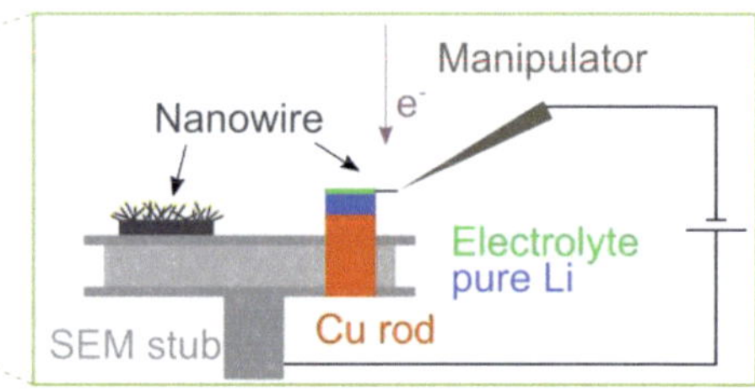


Figure A.2.1: In situ SEM nanowire battery setup. The optical image shows the setup configuration mounted inside the vacuum chamber. For better visualization the transfer system is opened under the air environment. The manipulator is connected to a galvanostat and subsequently single nanowires are harvested and dipped into the electrolyte with their distal end. By varying the potential charging/discharging is accomplished.

Once opened inside the chamber, single nanowires are harvested by the manipulator according to the procedure described in chapter 2.2.1. The manipulator is connected to a galvanostat/potentiostat (Keithley 2400 series source meter) where the microscope stage forms the other connection. The distal end of the single specimen is then pushed into the IL. Subsequently charging and discharging is accomplished by a variation of the voltage. During these processes, SEM images are recorded. After the testing is finished the nanowire is placed onto a TEM grid from where it can be transferred to the tensile testing unit for subsequent mechanical investigation.

## A.3   Gripping and cross sectional measurement

Gripping of individual nanowires is achieved using local deposition of the Pt-organic compound from the gas injection system of the microscope. In the case of small wires it is sufficient to place them on a surface and simply deposit a layer a couple of hundreds of nanometers thick. For thicker wires, roughly exceeding the diameter of 200 nm this may no longer be sufficient to guarantee proper gripping. However this also depends on the strength of the investigated material and the applied forces the Pt-nanowire and Pt surface interlaces have to withstand. In the case of larger wires Pt-organic will be deposited onto the wire and onto the surfaces that can be accessed from the top. The surfaces parallel to the e-beam cannot be reached and hence gripping will only be achieved once the layer of the surface and the wire coalesce.

Therefore a special technique for gripping larger specimens was developed. This special route is shown in Figure A.3.1. Prior to gripping a trough with dimensions slightly exceeding the dimensions of the specimen is cut into the surface of the bearing using FIB milling. Subsequently, the distal end of the wire is brought into the trough and deposition is started. The trough enables the Pt-organic compound to seep around the entire part of the specimen in the trough and therefore to provide the strongest grip possible.

Accurate measurements of cross sections are crucial to the accuracy of stresses, which are calculated by dividing the measured force by the nominal cross section. Therefore two different routes for the acquisition of cross sections have been developed, (i) bending the wire towards the e-beam with the manipulator and (ii) mounting the sample tilted by $90°$ so that the fracture surfaces of the wires are facing the e-beam. While the former is only applicable to ductile specimens, the latter can be used for both and is desired if the cross section is to be investigated without damaging the specimen. For ductile materials it is necessary to mill the fracture site to get a clean view on the nominal cross section, as shown in Figure A.3.1D.

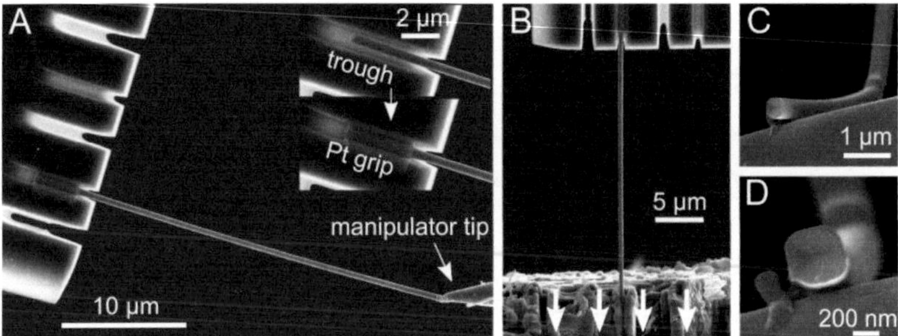


Figure A.3.1: Trough cutting and gripping technique for mounting of larger specimens. Trenches are FIB milled into both bearings, the silicon cantilever and the diamond cube of the transducer. The distal end of a fiber is brought into the trough with the manipulator and subsequently Pt-organic is deposited (A) and can seep around the whole diameter to provide the strongest grip possible, as shown in the insets. (B) shows the final test configuration. For post mortem analysis of cross sections ductile specimens are bent with the manipulator (C), so that the wire axis is parallel to the e-beam (D). FIB milling is employed to cut the fracture site enabling accurate measurements of the cross-sectional area.

## A.4    Weibull Analysis

Weibull analysis is commonly performed on a set of data having the same size (volume, surface area, length etc.), which allows for a straightforward Weibull analysis. However, the data pool of the nanowires does not exhibit a proper quantity of a certain volume etc. to perform reliable analysis and to make accurate predictions. Therefore a different route was developed by Dr. Martin Härtelt of the probabilistic group at KIT facilitating the incorporation of all data.

In addition to the cumulative distribution function (cdf) given in equation 5, the respective probability density function (pdf) is given as:

$$f = \frac{m}{b}\left(\frac{\sigma}{b}\right)^{m-1} exp\left[-\left(\frac{\sigma}{b}\right)^{m}\right] \tag{A.2}$$

The functional dependency of strength and the other parameter, e.g. the volume $V$ of the specimen is obtained from the weakest-link approach with $k=-1/m$:

$$b = \left(\frac{V}{C}\right)^{k} \tag{A.3}$$

Equation (A.3) inserted into (A.2) returns the new pdf with parameters $m$ and $C$ *(MPa·mm)*:

$$f = m\left(\frac{V}{C}\right)\sigma^{m-1}exp\left[-\sigma^{m}\left(\frac{V}{C}\right)\right] \tag{A.4}$$

Approximation of the parameters was done by means of the Maximum-Likelihood method with $N$ being the number of data sets of $\sigma_i$ and $V_i$.

$$LF = \prod_{i=1}^{N} f_i \rightarrow max \tag{A.5}$$

With this approach all acquired data can be employed irrespective of the specimen dimension to obtain reliable results. The calculated Weibull moduli are shown in Table A.1.

| Sample | $m_{surf}$ (measured) | $m_{surf}$ (corrected) | $m_{vol}$ (measured) | $m_{vol}$ (corrected) |
|---|---|---|---|---|
| Au nanowires | 3.4 | 5.6 | 3.7 | 6.5 |
| Au nanoribbons | 4.7 | 6.2 | 5.2 | 6.6 |
| Si grown on $SiO_2$ | 4.3 | 4.8 | 4.5 | 5.0 |
| Si grown on SOI | 7.2 | 11.5 | 7.8 | 11.8 |

Table A.1: Calculated Weibull moduli for various specimens

## A.5 Consecutive Testing of NaCl Nanowires

NaCl nanowires were tested in tension. After the first test one of the fractured portions was regripped and tested again (Figure A.5.1). Different tests are shown in different colors. In one case (Figure A.5.1A) a nanowire was regripped twice. All tests show linear-elastic loading and subsequent plastic deformation at similar stress levels. The amount of plastic deformation seemed to be random.

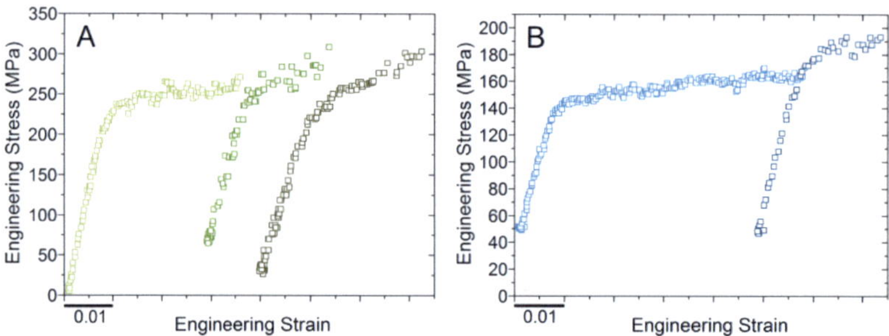


Figure A.5.1: Consecutive testing of NaCl nanowires. A fractured portion of a tested specimen was regripped and tested again. Linear-elastic loading is followed by plastic deformation at similar stresses and varying amounts of deformation.

## Schriftenreihe
## des Instituts für Angewandte Materialien

ISSN 2192-9963

Die Bände sind unter www.ksp.kit.edu als PDF frei verfügbar oder
als Druckausgabe bestellbar.

Band 1      Prachai Norajitra
**Divertor Development for a Future Fusion Power Plant.** 2012
ISBN 978-3-86644-738-7

Band 2      Jürgen Prokop
**Entwicklung von Spritzgießsonderverfahren zur Herstellung von
Mikrobauteilen durch galvanische Replikation.** 2011
ISBN 978-3-86644-755-4

Band 3      Theo Fett
**New contributions to R-curves and bridging stresses – Applications
of weight functions.** 2012
ISBN 978-3-86644-836-0

Band 4      Jérôme Acker
**Einfluss des Alkali/Niob-Verhältnisses und der Kupferdotierung auf
das Sinterverhalten, die Strukturbildung und die Mikrostruktur von
bleifreier Piezokeramik $(K_{0,5}Na_{0,5})NbO_3$.** 2012
ISBN 978-3-86644-867-4

Band 5      Holger Schwaab
**Nichtlineare Modellierung von Ferroelektrika unter Berücksich-
tigung der elektrischen Leitfähigkeit.** 2012
ISBN 978-3-86644-869-8

Band 6      Christian Dethloff
**Modeling of Helium Bubble Nucleation and Growth in Neutron
Irradiated RAFM Steels.** 2012
ISBN 978-3-86644-901-5

Band 7      Jens Reiser
**Duktilisierung von Wolfram. Synthese, Analyse und Charakterisierung
von Wolframlaminaten aus Wolframfolie.** 2012
ISBN 978-3-86644-902-2

Band 8      Andreas Sedlmayr
**Experimental Investigations of Deformation Pathways in Nanowires.**
2012
ISBN 978-3-86644-905-3